Already Extinct

A Parrhesiastic Statement

Minoru Kyo

Fescue Collective

ISBN 978-0-9880866-0-9

Published by Fescue Collective, Canada.

For information, visit http://www3.telus.net/fescuecollective/

Distributed by Booklocker.com, Inc., Bradenton, Florida. Also available in electronic format.

First Edition

Printed in the United States of America on acid-free paper.

Already Extinct

A Parrhesiastic Statement

Minoru Kyo

Fescue Collective

Dedicated to

the Koch brothers,

David and Charles,

who have done more than most

to make this book possible:

hupo philanthropias

Table of Contents

Figures

The Autistic Subject

Žižek suggests that there are three modes of response to an 'unheard-of experience' or trauma: the first (and rarest form of response) is to rationally reevaluate one's previous certainty; in the second, one becomes incapacitated from thinking and judging; and the third mode is denial. What trauma have we collectively experienced? Our trauma, it seems, has occurred at the three levels of our 'commons': the commons of external nature, of inner nature, and of symbolic substance. This 'unheard-of experience' is ultimately the failure and breakdown of the structure of meaning. But what do we really mean by 'meaning' and, more importantly, who really means by 'meaning'?

The state of our collective response to trauma can be informed by an understanding of the death-drive, and the death-drive can be used to inform us about the subject of this trauma. Indifferent and detached, the autistic subject is disengaged from the core of existence, void of substantive identity, devoid of symbolic texture: nothing remains but the survivor of its own death; nothing remains but the empty place of the subject. It is interesting that in the past thirty years, as a response to the trauma of neoconservative totalization, we have witnessed the emergence of postmodern thought, and it is interesting how this thought has been subverted by its transformation into the autistic subject.

The dyad of totality and its rupture describes the dialectic of existence. Take, for example, science and art. Both are processes of making meaning, but science does so by creating a totality – by taking the line of empirical evidence and turning it onto itself hermeneutically, to complete itself: a circle. Art, on the other hand, creates meaning that always exceeds meaning. It is a wellspring that continually overflows any attempt to contain it: the circle can never be closed – there is always an empty place that acts as the 'suture' for making meaning. This is as it must be. Science that totalizes becomes moribund without the rupture of art; and art without the structure of science is the impossible-Real: it is fear.

Similarly, postmodernism began as a wellspring that would rupture the totalization of Western thought manifested by the totalitarianism of neoconservative globalization and control. Philosophers found empty signifiers acting as sutures on these totalizations and tore them open (deconstruction) to reveal the deferral of signification and the always-overcoming of meaning. But for many, this proved to be a trauma too great to manage and, rather than revelation, the result was the autistic subject; rather than revolution, the result was despondence in the face of meaninglessness.

Words mean by signifying something. But signification never coincides with the thing-in-itself. Signifiers can be replaced and meaning deferred in a series without end (or beginning). Ultimately, the subject of the signifier coincides with its own impossibility, it is the void opened up by the failure of representation. Badiou might describe this as this empty set that is the subset of any set of representation – every array of numbers contains the empty subset. This empty subset is the nothing, the void, which is both representation and the failure of representation: it is the suture that bars totalization.

Žižek uses the example of value. In commodity transaction, one object can be traded for another, which can then be traded for a third object. The series of trades (as the representation of value) is infinite, and the meaning of value continues to overflow itself. In the transition from trading use-value to trading exchange-value, money becomes the empty signifier used to suture the meaning of the commodity, the coincidence of the proletariat and capital. Money is, however, an empty signifier – its value is not intrinsic to the thing-in-itself, it is a trust that makes the system of exchange function. In other words, money is the suture for the totalization of the meaning of value, it is the void that coincides with its own impossibility.

The point of signification becomes the extreme of its own self-negation. It is something reflected by its own absence; it is the suture of totalization that masks the rupture, or overflowing, of meaning; it is the Nothing that is represented by the One even in the possibility of its own absence. Žižek states:

> That is to say, as to its form, Being already possesses a determination of "something", yet its content is "nothing" – it is therefore "nothing in the form of something", Nothing counted as Something. Without this absolute tension, Being and Nothing would coincide immediately and the dialectical process would not be "set in motion".[1]

Being itself is the result of this dialectical dyad, it is the virtual point of its self-relating. It is constituted by means of a double reflection in which the past is mirrored in the future to reflect the present which has no past. It retroactively posits the conditions of its necessity. It is the iteration of the unsayable. There is no originary subject, no totalized being. This failure to totalize allows the overflowing of being which is reflected back from an outside-of-totality, a Beyond that does not exist. To be clear:

> It is not enough, here, just to draw attention to the elementary fact that, with Hegel, judgment is not "subjective" in the customary meaning of the term but a matter of the relationship of the object itself to its own

> Notion – the radical conclusion to be drawn that *there is no Subject without a gap separating the object from its Notion* – that this gap between the object and its Notion is the ontological condition of the Subject's emergence. The Subject is nothing but the gap in the Substance, the inadequacy of the Substance to itself: what we call "Subject" is the perspective illusion by means of which the Substance perceives itself in distorted ("subjective") form. More crucially, the fact is here generally overlooked that such a type of judgment on the correspondence of an object to its own Notion implies a kind of reflective *redoubling* of the Subject's will and desire.[2]

The suture for the subject, the empty set, the articulation of the void is nothing other than the death-drive. The death-drive is marked by an impossible excess, it is the impossible moment of the birth of subjectivity; it is the giving-form of death in life. The death-drive is the tain of the mirror reflecting the present subject back from the future. It resists symbolization as it presents the prohibition of a beyond, the desire of the nothing that sutures the totalization of subjectivity, it is the gap of the Real that is both desired but cannot be faced. The death-drive is the traumatic core of being.

Consider the current state of being once the barriers to pleasure and enjoyment (*jouissance*) are removed. Without prohibition, enjoyment is prohibited: "*enjoyment itself, which we experience as 'transgression', is in its innermost status something imposed, ordered*"[3]. As the suture is detached, we face the trauma of the impossible-real.

> Dis-attachment is thus the death drive at its purest , the gesture of ontological 'derailment' which throws the

> order of Being 'out of joint', the gesture of dis-investment, of 'contraction'/withdrawal from being immersed in the world, and primordial attachment is the counter-move to this negative gesture.[4]

The condition described is the autistic subject, the 'undead' – already dead while living. Facing the 'unheard-of experience' of desuturing the death-drive from the totalizing of being, meaning appears manifested in the three modes of response to a trauma, including disorientation, detachment and denial. Can the autistic subject recover from this trauma? We'll see, but I doubt it.

The Trauma of External Nature

Oh, you're still reading!? I didn't expect it. I was under the impression that people couldn't stand anti-essentialist gobbledygook. The last chapter is the first reason that we are in trouble - we cannot even find a language or a common ground to communicate ideas anymore … the trauma of 'symbolic substance'. It only gets worse, you know, as we will deal with the trauma of 'external' and 'internal' nature. Maybe, though, you are of the type of person that deals with trauma by 'reevaluating your previous certainties' by challenging your ideologies (?) If not, this is no place for you …

Belated Preface

I will (really) begin with an explanation of the structure of the book that follows. I've learned that most people are not greatly influenced by facts. No one has ever changed their behavior because of a simple fact that they might have tripped across while scrambling through their daily existence. Most people already know what they want to know. On the other hand, most people desire the credibility that facts proffer to substantiate a person's perspective or interpretation of the real. So, what to do? Since I have no desire to change your behavior (on the contrary, I will be arguing that there is no point), I will simply state the case. If, in the course of reading this book, your certainties are challenged, do more research on your own. There was a mystic named Gurdjieff who argued that 'without struggle, there is no progress and no result' - I think this position is quite appropriate as it relates to thinking.

The book will outline our current sources of trauma - environmental, social, and economic. The general model, following Herman Daly, is that the economy is contained by society, and both are contained by the environment: one nestled in another like three Russian matryoshka dolls. As such, the topics cannot easily be disentangled at times. The complex interrelationship between environment, society and economy is what has largely been ignored in contemporary thought (particularly orthodox economics) and this is, perhaps, the source of our collective trauma - the failure and breakdown of the structure of meaning (as it relates to the commons of external nature, internal nature, and symbolic substance).

Shamelessly, I will be making a hostage of this book and tie it to the rock of a reductive thesis ... the argument will be: *we don't understand the problem; we couldn't do anything about it if we did; and we wouldn't do anything about it if we could.*

"And this nihilism found its perfect expression in the odyssey to the moon - because we went there without knowing why we went."

Norman Mailer

We don't understand the problem ...

The Trauma of Environment

Before we go too far, let's be clear: When we talk about the environment, we are really talking about how the environment serves us as a source of resources and as a sink for our wastes. There is, by contrast, a deep-green point of view for which the environment has a value intrinsic to itself. But if we cannot even value the environment enough to save our own skins, it seems a little farfetched that we will start 'thinking like a mountain' anytime soon, as Aldo Leopold has suggested we should. The environment (as a source of substances and sink for our wastes) will be divided into four major categories for our discussion: energy, mineral resources, soil and air. Let's begin with some thoughts on how we collectively impact the natural environment.

Sustainability Blah Blah

"If I had an enemy bigger than my apathy
I could have won."

Mumford and Sons

Unlike some other authors, I will not suggest that the earth will be swallowed by the sun in a few billion years, or that 99.999% of species that have existed on earth are already extinct. I will not, in these or any other ways, try to obscure the fact that our current situation is caused by our own actions, and that this current situation is not due to some natural or inevitable event. Our future is the result of what we have done in the past, and what we continue to do today - plain and simple.

Human life has an impact on the natural environment. Everything that exists has an impact on the environment, as it is part of the environment. We will talk about our collective impact in a moment, but it should also be understood that there is no 'outside' the environment. We are part of a natural system that is governed, amongst other things, by the Second Law of Thermodynamics - entropy - which describes the slow winding down of the clock that was last wound at the beginning of what we perceive to be linear time. Entropy is an unavoidable reality of our place in the environment. Kirschenbaum has expressed this physical reality as it relates to our interaction with the world:

> The loss of natural forest cover or its replacement with monocrop plantations results in simplification of ecosystems - entropy. The conversion of semiarid woodlands to desert through overexploitation results in ecosystem simplification - entropy. The erosion of topsoil results in diffusion of nutrients - entropy. The eutrophication of aquatic and marine environments from the diffusion of nutrients results in decreased biotic diversity and ecosystem simplification - entropy. The depletion of the world's fisheries results in ecosystem simplification - entropy. The loss of global biodiversity results in simplification - entropy. Global climate change due to the buildup of carbon dioxide in the atmosphere from the burning of fossil fuels in a process of diffusion of carbon - entropy.[5]

All systems are winding down to the lowest energy level or the simplest state of existence. At the surface of things, it appears that our impact within earth systems is part of an inevitable trend. At the surface of things we could say that our impact is natural. What is truly amazing, however, is that life in the biosphere works against this trend (though, overall, the law of entropy prevails). Within the biosphere, plants and animals adapt and rejuvenate as they interrelate within a changing environment. Life has become more complex over the eons as nature has created the dynamic and symbiotic environment in which we now live. What we currently describe as *impact*, then, is not the conversion of matter to sustain life - eating, building, travelling - but the *rate* of this conversion process. What we will describe as impact (as it relates to sustainability) is consuming resources at a pace that exceeds the rate of natural replenishment and making waste that exceeds the rate of absorption by natural ecosystems. Impact, therefore, does not describe our place in the natural order, it describes a rate of conversion of resources that puts the ecosystem out of balance, or weakens it to a point where it is overcome by the grim grasp of entropy.

In a similar way, the magnitude of your bank account has no impact on the sustainability of your expenditures: no matter what the size of the account, if you spend at a rate that exceeds replenishment (wages, interest, profits from speculation), the bank account will eventually be spent and you will be broke. Entropy is like inflation, which is the slow devaluation of your money in the bank. In a healthy economy (economic ecosystem) you might earn interest to counteract the inflation, maintaining a balanced account. So, if you spend at the rate your bank account is replenished, you can expect to have money indefinitely and the original capital can be passed on from generation to generation - the accumulated social wealth saved for a sustainable future.

Now, consider an earth system, which is a system of organic and inorganic parts that interact to maintain a complex balance. In this system of parts, the balance is created when each part can access what it requires, and its waste or output is absorbed as a resource for another part of the system. Each part has no net impact: it is a natural system in sustainable balance. In such a system, there is a mutualism between the parts which can be sustained indefinitely (as long as the sun keeps shining). On the other hand, a system in which one component limits or diminishes the ability other components to be perpetuated within the system describes an impact. In such a system, one part benefits at the expense or even the detriment of another part, characterizing a form of parasitism that cannot be sustained.

Similarly, in describing an economic system, a person receives a wage for their labour power, and spends this wage to sustain their individual self and their family (so as to reproduce the workforce). From a purely economic perspective (ignoring for now the environmental impact of commodity production, the psychological impact of drudgery, and the alienation of the worker from the fruits of their labour), this person is part of a system of production and reproduction that can be sustained in perpetuity. The price realized for the commodity or service is used to reproduce the workforce and replace the capital used up in production. Money circulates in an endless circle. People produce what they need (food, clothing, shelter, books) to reproduce the species indefinitely. This is a steady-state economy, of sorts.

If, however, the price realized for the commodity or service is used to pay wages and replace capital as before, but this time with some money (mysteriously) left over. This profit can be reinvested to expand the production of commodities. More production means more workers and more workers means more product, and more product means more consumption, and more consumption means more profits ... which must be reinvested in an ever-expanding cycle. In such a system, one part (the part that controls the profits) benefits at the expense of the other (who receives the wages). The increasing rate of commodity production is limited by investment opportunities that will result in future profits, which is in turn limited by the growing disequilibrium of money circulating within the system (a disequilibrium between those who have too much and those who have too little). The increasing rate of commodity production also increases the rate of extraction of resources from the environment, and the inevitable discharge of waste into the environment. This describes a system that cannot be sustained. When one considers the ever-growing material inputs (from the earth and from living ecosystems) and wastes generated by an ever-expanding system of commodity production, and when one considers the ever-growing population of workers required to produce and consume these commodities, the impetuses behind a growing net human impact become evident: an economy premised on perpetual growth is not sustainable.

I = PLOT

What are the dimensions of human impact on the earth? In the 1970s, a relationship between population, affluence and technology was developed as three factors of human impact. This relationship was given the equation I (Impact) = P (population) x A (affluence) x T (technology). In effect, it was posited, we are having an impact on the earth because there are too many of us, we want too much, and the technologies we develop to sustain the system are becoming too complex and powerful. There was one thing that was missing from this relationship, an unknown-known, a blind spot. Realizing this, Patrick Curry[6] has suggested a fourth dimension: Organization. The dimension of Organization represents the way humans structure their societies: the institutions and ideologies that inform their economic systems, forms of government, and religions. Curry reframed the equation to become I = PLOT:

Human Impact
= Population x Lifestyle x Organization x Technology

It is important to note that each of these dimensions simultaneously affects human impact on the earth - they can act as independent effects, and they can interact in positive or negative feedback loops. There is always some confusion around concept of feedback: a positive feedback reinforces and adds to the original direction, and a negative feedback resists or corrects the original direction. An example of a positive feedback is snow in winter. As the temperature drops in winter (when the earth is tilted away from the sun), it begins to snow. The snow reflects the sun's radiation making it even colder, which results in yet more snow. Another example is global warming due to greenhouse gases in the atmosphere. As the average global temperature rises due to greenhouse gas emissions, the permafrost in the arctic begins to melt which releases vast quantities of methane (a potent greenhouse gas), and this, in turn, increases the rate of global warming (and the rate of melting of permafrost). The result is negative for life on earth, but it is a positive feedback mechanism. An example of a negative feedback is population growth, in which an increasing population of an animal in an ecosystem is faced with a reduced amount of available food per capita (unless the food source also increases). Without enough food, the population begins to decline. A negative feedback mechanism tends to work against a trend, often to correct it back towards a balance.

Reducing human impact will depend on the ability to control the dimensions of Population, Lifestyle, Organization, and Technology, and it is my intention to show that each dimension is largely inelastic in the direction that they can be controlled. A brief introduction to each of these dimensions will be followed by some of the current manifestations of human impact.

The human Population has reached 7 billion persons and is expected to exceed 9 billion by mid-century according to United Nations projections. It took millennia to reach a population of 3 billion people on earth, and it has taken only the last fifty years to double this number. We are presently adding a small city on earth each and every day. Most of this population growth is accruing in developing nations, whereas more developed nations are currently relatively stable in population, as shown. (Before you jump into the blame-game against developing nations, remember population is only one of the dimensions of Impact ...)

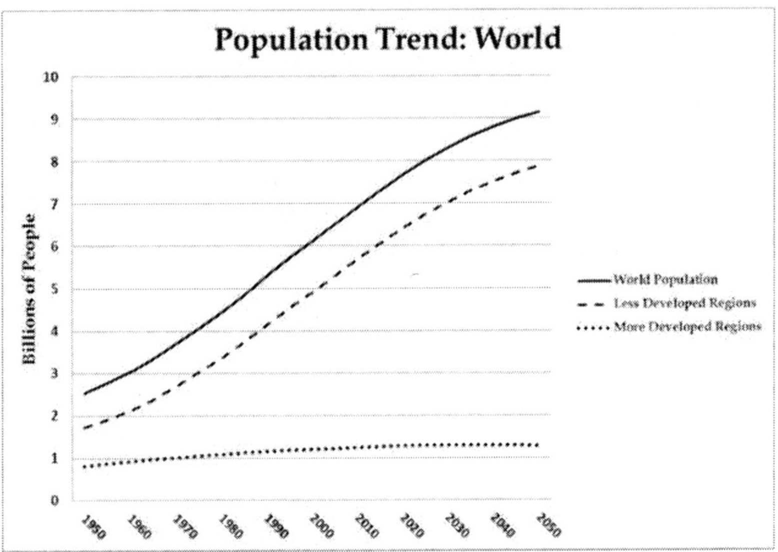

Figure 1 - World Population Growth (derived from United Nations data)

The human population requires resources to feed and sustain itself (arable land and fresh water), a stable climate (for timely water availability and natural processes of plant growth and fertilization), and materials to provide shelter. If these resources are available at a minimum level for all people, arguably the earth can sustain a large human population; some have suggested the earth has a carrying capacity (at minimum levels of subsistence) of 17 billion people. But if this population expects a higher level of wellbeing (Lifestyle), then only a much smaller population is sustainable, perhaps only one or two billion people at a lifestyle typical of the average North American. Population control and population reduction are contentious issues, and they are generally avoided by people trying to deal with sustainability concerns – the preference is to talk about lifestyle as the main culprit, or technology as the principal redeemer.

Lifestyle describes the quality of life or well-being of a population, and it also describes the consumption (or overconsumption) of commodities by a population. The impact of lifestyle is closely related to population from a human impact perspective. The following chart shows the considerably larger growth of human populations in both China and India compared to the United States. Based solely on population growth, one might argue that these burgeoning populations must be controlled to reduce the human impact on the planet.

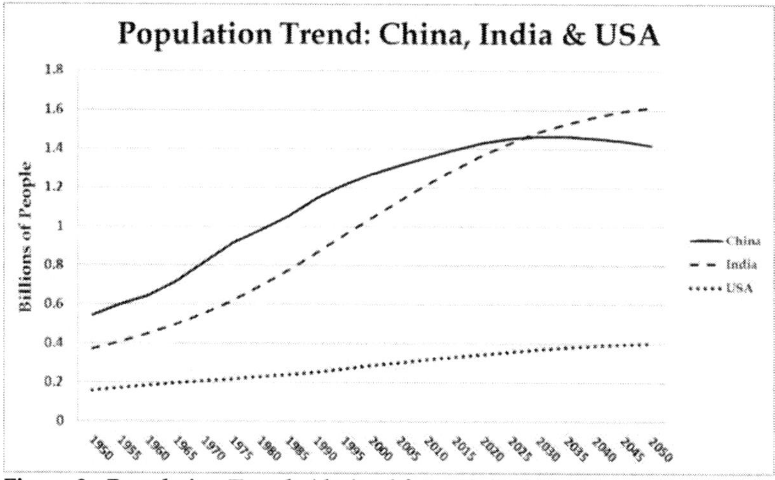

Figure 2 - Population Trends (derived from United Nations data)

But if one were to address the Ecological Footprint for each, a different story emerges. The Ecological Footprint describes the amount of area (land and sea) required to yield the primary products consumed by a population, including cropland, grazing land, forests and fisheries. The United States consumes roughly four times the primary products per capita than China, and eight times that of India. This may be compared to biocapacity, which is the amount of productive land and sea available to provide these products. Trends show that the ecological footprints of all three countries exceed their regional biocapacity.

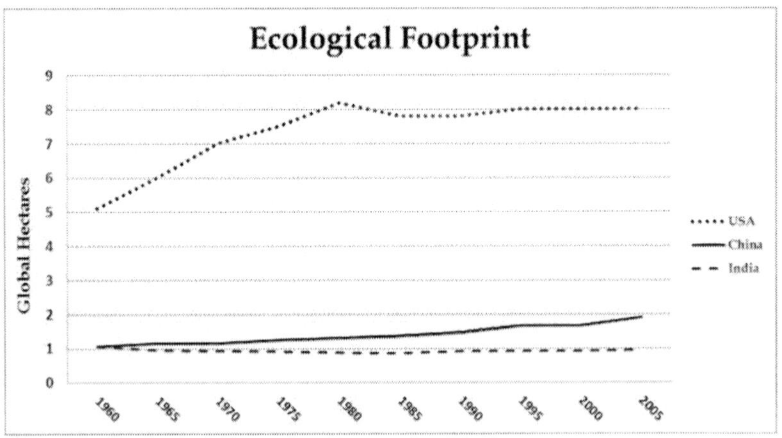

Figure 3 - Ecological Footprint Comparing Nations

There are other dimensions to measure the impacts of lifestyle. Using just the Ecological Footprint, every person born in the United States over the next forty years will have an impact that is eight times more than that of every person born in India. Based on expected population trends, there will be 4.6 people born in India for every birth in the United States. In this example, it is clear that each birth in the United States has a greater impact than a birth in India. So, at current levels of lifestyle, when one speaks of population control one should be talking about controlling affluent nations like the United States. Maybe that's why the subject is contentious: imagine the United States being asked by the world to enact a one-child policy.

As discussed, Organization as a dimension of human impact represents the way humans structure their societies: the institutions and ideologies that inform their economic systems, forms of government, and religions. These are the institutions and ideologies that support human exceptionalism (that we are more important than other species); that support human entitlement to affluent lifestyles; that promote population growth; and that support infinite economic growth. As such, our institutions and ideologies greatly influence the magnitude of human impact. Organization also allows human populations to live and prosper in large urban centers with the potential of optimizing resources for growing human populations. In general, the effects of organization could be subsumed under the other dimensions of human impact (population, lifestyle and technology), but organization profoundly influences the *elasticity and inertia* of change and, as such, merits its own dimension. To be clear, without organizational and institutional change, none of the other factors of Impact can be managed for positive change. I will focus later mainly on the impotence of government and the perversity of capitalism to illustrate this.

And, finally, Technology as a dimension of human impact has historically resulted in contradictory affects. Technology can be harnessed to reduce human impact by increasing the yields of primary products, by improving the efficiency of the use of energy and raw materials, by providing for pollution abatement (air, water, and solid waste), and by enhancing the potential for a healthy and educated population. Technology, however, has also been blemished by ever greater destructive power, by greater complexity, and by accelerating the creation of pollution with increased energy and commodity production.

The interrelatedness of population, lifestyle, organization and technology is clear - no single dimension is absolutely responsible for human impact, and no single dimension will ensure our ability to achieve sustainability within the regenerative capacity of the earth. Technology can sustainably enable greater human population and improved lifestyle, and lifestyle can be sustainably improved with better organization which might lead to a reduced birth rate, and better human organization might allow for reduced lifestyle expectations and a more effective deployment of necessary technologies. But once we have optimized each of these dimensions, we will have reached the limit of our collective ability to sustainably maintain a human population on earth.

What is this limit? Where is the balance? Let's begin to answer these questions by evaluating the current state of the earth, and our appetite for resources.

The Colossal Use of … Everything

As a dimension of human impact, it is important to understand the current state of the earth as it relates to our consumption of resources and the disposal of our wastes to the air, water and soil around us. The Global Footprint Network suggests that our current Ecological Footprint exceeds the earth's capacity by 50% - that is, we need one and one-half planet earths to sustain our current consumption of primary products. This global ecological footprint has increased at a rate greater than the increase in population, as lifestyle expectations have rapidly grown. In the United States, consumption increased sixfold since 1960 even though the population had only doubled. The consumption per person, therefore, almost tripled in the last fifty years. Furthermore, over the same period, "metals production grew sixfold, oil consumption eightfold, and natural gas consumption 14 fold. In total, 60 billion tons of resources are now extracted annually - about 50 percent more than just 30 years ago."[7]

> The bottom line is different from that generally assumed by our leading economists and public philosophers. They have mostly ignored the numbers that count. Consider that with the global population past six billion and on its way to eight billion or more by mid-century, per-capita fresh water and arable land are descending to levels resource experts agree are risky. The ecological footprint - the average amount of productive land and shallow sea appropriated by each person in bits and pieces from around the world for food, water, housing, energy, transportation, commerce, and waste absorption - is about one hectare (2.5 acres)

> in developing nations but about 9.6 hectares (24 acres) in the United States. The footprint for the total human population is 2.1 hectares (5.2 acres). For every person in the world to reach present U.S. levels of consumption with existing technology would require four more planet Earths. The five billion people of the developing countries may never wish to attain this level of profligacy. But in trying to achieve at least a decent standard of living, they have joined the industrial world in erasing the last of the natural environments.[8]

Clearly, the ecological footprint is not distributed evenly over the population, neither within the United States nor from a global perspective. Annie Leonard in *The Story of Stuff* describes this as follows:

> The richest fifth of the global population consumes 45 percent of all meat and fish; the poorest fifth 5 percent. The richest fifth consumes 58 percent of energy generated globally; the poorest fifth less than 4 percent. ... The richest fifth consumes 84 percent of all paper; the poorest fifth 1.1 percent. The richest fifth owns 87 percent of the world's vehicle fleet; the poorest fifth less than 1 percent.[9]

It is plain to see that we consume a lot of resources, and the richer we become, the more we consume. Unfortunately, at a certain point the benefits of expanding consumption begin to level off. For many of the world's poorest, the consumption of resources is related to meeting or just exceeding basic needs – an increase in consumption means a direct increase in well-being for this group of people. Researchers have suggested that the correlation between consumption and well-being is reached with a per capita expenditure of $15,000 to $20,000 (in current purchasing power)[10]. Beyond this level, it seems, the satisfaction derived from consumption is much less, and may even reach a point where it has a negative impact on well-being (taking into consideration, for example, the stress caused in the effort to maintain the cash flow required to maintain an expected lifestyle).

So, why do we continue to consume beyond what is truly satisfying? Some believe that we are being programmed to consume everything that is within our grasp, and others submit that we must use up all of the stuff that we create to keep the economy going. Leonard suggests that after World War II the productive capacity of the economy had to be redirected in order to keep the factories producing, and this was accomplished by keeping the people consuming. She quotes the analyst, Victor Lebow, as saying: "Our enormously productive economy … demands that we make consumption our way of life, that we convert the buying and use of goods into rituals, that we seek our spiritual satisfaction, or ego satisfaction, in consumption … we need things consumed, burned up, replaced and discarded at an ever-accelerating rate."[11] Veblen has labeled this as 'unproductive consumption' – that is, consumption that satisfies no real need except the creation of culturally specific impressions. As such, we have fashioned a society that has found 'value' in consuming things they don't need, and these values can only be described as manufactured. Veblen remarks: "Unproductive consumption of goods is honourable, primarily as a mark of prowess and a perquisite of human dignity; secondarily it becomes substantially honourable in itself, especially the consumption of the more desirable things."[12] In other words, the vast (unproductive) consumption of finite natural resources and the generation of wastes that must be absorbed by the environment is premised on Organization - on manufactured values and expectations.

To summarize, the global ecological footprint already exceeds the capacity of the planet, and much of this excess is related to unproductive consumption by a small, but wealthy, group of individuals who derive little happiness or well-being from this consumption. Reducing the human impact from unproductive consumption will require organizational change (not to mention population control and beneficial technologies). And this isn't going to happen overnight or, perhaps at all (as will be discussed later). In the meantime, we are relying on a one-time inheritance of fossil fuel energy, arable land, clean lakes and streams, biodiverse forests and plains, as well as a climatic balance spanning the period of human civilization.

> An international analysis of the world's ecosystems, written by more than 1,300 scientists, found that 60 percent of the services of nature - including those provided by farmlands, fisheries, and forests - are being degraded or used unsustainably. The Millennium Ecosystem Assessment warned in 2005 that "these problems unless addressed, will substantially diminish the benefits that future generations obtain from ecosystems.[13]

We have already defined *impact* as the consumption of resources at a pace that exceeds the rate of natural replenishment and the production of waste that exceeds the rate of absorption by natural ecosystems. It will be argued that the human community is presently consuming resources in excess of the rate of replenishment, and it is creating waste at rates that vastly exceed the ability of the earth to absorb. But it's not our fault, is it? I mean, it is not our *intention* to consume at a rate that threatens the future of human civilization, is it? That reminds me of a quote by Tolstoy who wrote in 1886: "I sit on a man's back, choking him, and making him carry me, and yet assure myself and others that I am very sorry for him and wish to ease his lot by any means possible - except getting off his back." Tolstoy was writing about the unsustainable institution of serfdom, but it could just as easily describe our present unsustainable relationship to the earth: we'll do anything to reduce our impact on the earth … except change.

Fossil Fuel Forever!

What resources are we consuming unsustainably? What inheritances are we squandering? The first thing that should be considered is fossil fuel energy, as energy is perhaps the most important and ubiquitous resource contributing to our lifestyle. Of our current consumption of primary sources of energy in the world, almost 90% are fossil fuels: coal, oil and natural gas. The sources for the remaining 10% include nuclear energy, hydroelectricity and renewables including biomass, solar, tidal, geothermal, and wind.

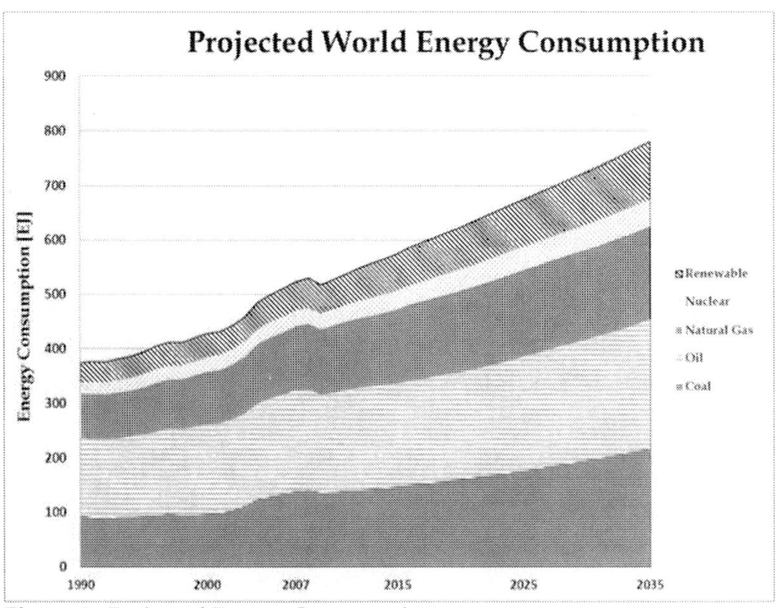

Figure 4 - Projected Energy Consumption

Based on a business-as-usual perspective, the projected consumption of energy will continue unabated well into the future, and it is anticipated that there will be little change in the relative importance of fossil fuels as a primary source of energy. It should be noted that fossil fuel energy sources are a one-time inheritance of energy - they were created over many millions of years and the rate of replenishment is practically nil. As such, we should know that these projections are not realizable in the longer term. It should also be noted that the growth of energy consumption over the next twenty years is more than double the forecasted capacity of renewable energy sources.

Today we consume about 475 exajoules of energy each year. If all that energy was to come from oil, that would be 79 billion barrels of oil equivalent - that's 3.3 trillion gallons each year! With that, you could fill the gas tank of 10,000 cars each second, continuously. Or, you could fill the Empire State Building 32 times each day. I don't know why you would want to do that, but this is clearly a lot of energy, and most of it is being withdrawn from our (one time) energy-inheritance. (As a side note, I have put these and the many numbers to follow in an Appendix at the back of the book - for those of you who like to check figures, you can see where mine were derived).

To put this in perspective, each person can generate about 100 W of power continuously over a working day. This number is based on a healthy diet of 2500 calories each day. So, a person over an eight-hour day can do 2,880,000 joules of work. If we took the entire amount of primary energy we use each year and convert it to work at a low conversion efficiency of 30% (meaning that only about a third of the energy is converted into work - the rest being lost as waste heat), it is like having 135,600,000,000 people doing free work for us each day to benefit our comfortable lifestyle. That amounts to almost

20 energy-people for each real person on earth. These energy-people do not have to be fed or housed, thank goodness.

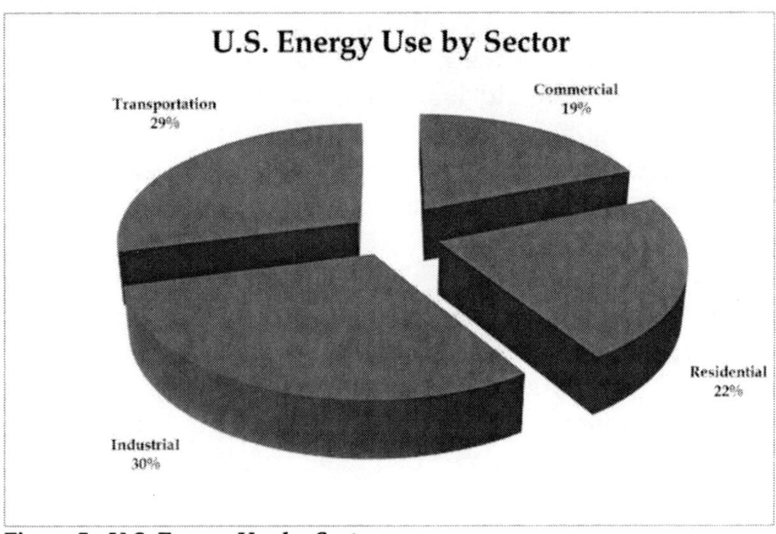

Figure 5 - U.S. Energy Use by Sector

That's the big picture of energy, and the point is: we use a lot of it. The energy produced from sources of coal, oil, and natural gas is consumed, however, in different ways in different amounts in different countries. Not all of the coal, oil or gas is burned for its energy, as much of it is converted into durable goods (like plastics) or used to make fertilizers, herbicides and pesticides for agriculture (and our luxurious lawns). For the coal, oil and gas actually burned for energy, using the United States as an example, the distribution can be broken into the following sectors: transportation, industrial, commercial and residential. It is said that about half of the energy consumed each year goes to the heating and cooling of buildings (industrial, commercial, and residential). Another third is used for transportation, including personal travel and the movement of goods around the globe in our efforts to access impoverished labour markets and exploit minerals from areas with lax environmental regulations (what is termed by economists as 'comparative advantage').

Oil

Oil is the primary energy used for transportation, and it is a significant source of energy for electricity generation and industrial processes. The world production of oil is shown using data from British Petroleum[14]. In 2009, the world was producing almost 1000 barrels each second.

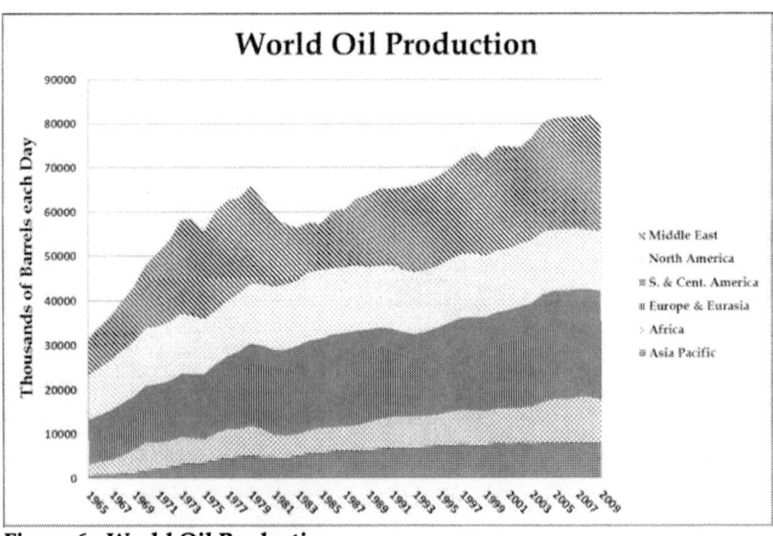

Figure 6 - World Oil Production

Note the decline of almost a million barrels a day in 2007, preceded by a plateau of four or five years. You will note the same decline in the previous figure on Projected World Energy Consumption, which is followed by a long-term projected increase in energy demand. A growing number of people, however, accept the principle that the rate of the world production of oil has peaked - that the long-term projected increase in oil production (shown in the previous figure) will not be realized. The following chart, based on data from Ziegler, Campbell, & Zagar,[15] shows a forecast of oil and gas rate of production based on known reserves and a realistic performance for recovering these resources using current technologies.

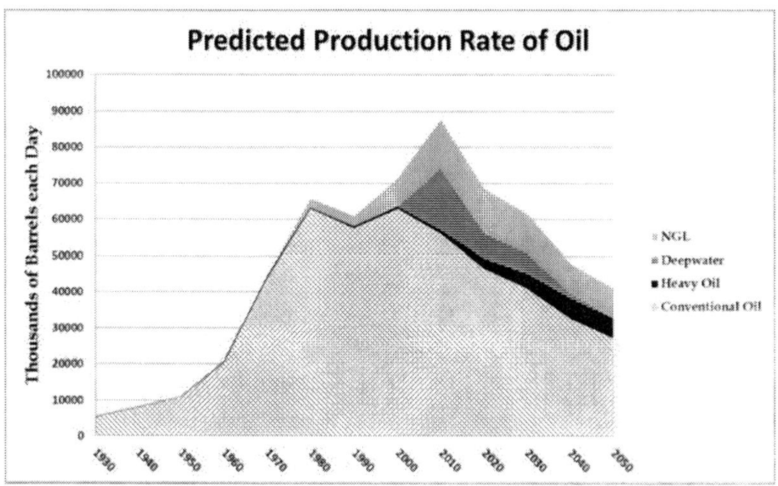

Figure 7 - Peak Oil

The production of oil in 2007 corresponds with the 80 million barrels of oil per day production rate (almost 1000 barrels per second), as shown in the previous figures. The predicted rate of production curve differs from these other charts in that it shows a maximum production rate roughly beginning in 2007 followed by a steady decline thereafter. This is called Hubbert's Peak named after an American geophysicist who, in the 1950s, predicted that oil production would peak in the United States in the 1970s … which it did.

Peak oil does not mean that the earth has run out of oil reserves. The misunderstanding that peak oil means that we have run out is the usual response from people suffering from the trauma of being presented an idea that they wish not to believe. No. There is a lot of oil still in the ground. What peak oil means is that oil cannot be produced at ever-increasing rates - at a certain point, the rate of production reaches a maximum and then declines. This has been demonstrated by single wells, by oil fields, by whole countries, and it is currently being demonstrated at the global level. The reason for peak oil is that the rate at which an oil field, or even an individual oil well, can be produced is limited by the geology of the formation. Oil is found in a sandstone-like rock that allows the oil to flow at rates that depend on the porosity (sieve-ness) of the rock and the viscosity (runniness) of the oil.

In the exploitation process, a well is drilled from the surface through many layers of sedimentary rock and water aquifers, until the hole reaches the layer with the oil or gas in it. The depth of the hole can range from a few hundred meters to many kilometers. Once drilled and completed, the oil nearest to the wellbore is produced relatively easily, while the oil from further away takes longer to reach the wellbore to be produced. Due to geological factors, the oil production rate will approximate a bell curve. By way of comparison, consider a

Slurpee – a crushed-ice drink filled with sweet syrup. When you buy the drink you might sip on it slowly at first to avoid brain-freeze. Not long afterward, your rate of drinking accelerates until it reaches a maximum. This maximum is limited by the rate at which the sweet syrup can move through the ice to the straw, not to mention the rate at which your body desires the liquid (the demand for Slurpee). You may sustain this peak for a period of time, but eventually the rate of drinking will begin to decline as you wait for the syrup to travel from the farthest reaches of the ice in the cup to your straw. By the end of the productive life of your Slurpee, you are drinking at the rate at which the ice is melting. Sometimes you abandon the Slurpee at this point because it is not worth waiting for. To take the metaphor one step further, in real life you might at this point toss the cup onto the ground, expecting someone else to clean it up for you – privatize the drink, socialize the garbage. (As a side-comment, Alberta – an oil-producing province in Canada – has tens of thousands of wells that have been abandoned and not reclaimed, with a total liability of $21 billion for reclaiming the impacts of existing oil & gas infrastructure.)[16]

As more wells are drilled into the field, the combination of wells will follow a similar curve. As each well draws from a large area surrounding the wellbore, drilling more wells at closer spacing will result in less oil produced by each (they are competing for the available oil in the formation). To illustrate this, imagine trying to empty a barrel of apples. One person can empty the barrel in 100 minutes. A second person might double this rate and empty the barrel in 50 minutes. Four people might even be able to empty it in 25 minutes. But adding a fifth person would not lower it to 20 minutes as the extrapolation suggests, but only to 22 minutes. Ten people might only be able to empty the barrel of apples in 20 minutes.

Why can't ten people empty the barrel in 10 minutes? At a certain point, the people are getting in each other's way, and some cannot even reach the barrel. Similarly, for an oil or gas field, there is a point at which adding an additional well will have only a marginal benefit in the effort to increase production - each oil field has limits as to how much of the resource it will yield.

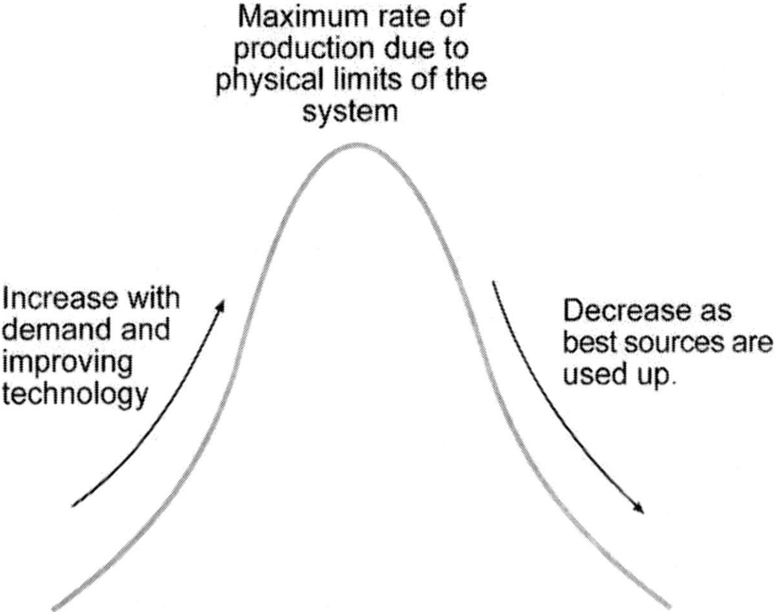

Figure 8 - Peaking Energy Resources

At a national and world level, the oil reserves that are large, light (high viscosity), and sweet (without contaminants), and easy to access have been exploited first. We are now exploiting reserves that are much smaller, heavier (as heavy as the Alberta tar sands), sour (with sulphur), and found in locations that are more difficult to access. As such, the world production will follow a bell curve with a growth period tracing demand, a plateau or peak, and an inevitable decline. It is kind of like running down a steep hill. You begin under control and pick up speed quite quickly. At a certain point you are running as fast as you can, being limited by your physical ability to move your legs quickly enough. Beyond this peak, you inevitably fall on your face and tumble down the hill until you stop at the bottom. The point of peak oil that that the rate of production will not be able to keep up with the rate at which we wish to consume the energy - and this disparity will have profound effects on our societies.

Minoru Kyo

"The oil industry and our oil-based economy, not just in Canada but in the world, depend on two things for their continued existence. The ability of geologists to find new fields of oil and our willingness to ignore the obvious, that, at some point, we're going to run out of oil. This would suggest that reducing energy consumption, curbing the proliferation of private cars and multilane highways, and converting to sustainable and reliable sources of energy such as solar, tidal surge, or wind power would be our first priorities. In fact, we have no such priorities. We have only the hope that the exhaustion of the oil supply will not happen in our lifetime.

It's not that we don't care about ethics or ethical behaviour. It's not that we don't care about the environment, about society, about morality. It's just that we care more about our comfort and the things that make us comfortable - property, prestige, power, appearance, safety. And the things that insulate us from the vicissitudes of life. Money, for instance."[17]

Thomas King

Natural Gas

The production rate of natural gas is similarly restricted by the size of the discovery, the porosity of the rock, and the location of the discovery. Unlike oil, natural gas is difficult to transport overseas (as it needs to be cooled and compressed into a liquid), so domestic consumption tends to drive production. As shown in the following figure, the rate of conventional gas production is also peaking, and will begin to decline, even with non-conventional sources being rapidly exploited to fill the gap.

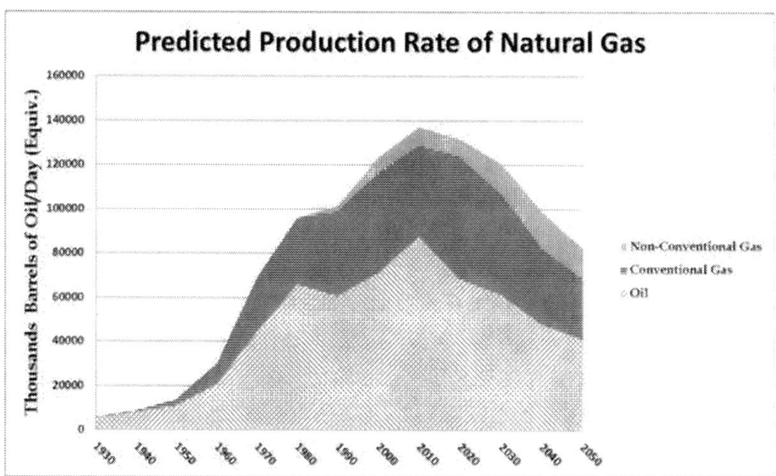

Figure 9 - Peak Natural Gas

Non-conventional sources of natural gas such as coal bed methane or shale gas are being developed using advanced drilling and fracturing technologies. These technologies involve drilling long horizontal wells in the formation and using complex (and toxic) fluids under high pressures to break up the gas-bearing strata to allow the gas to more freely flow to the well. Unfortunately, many of these non-conventional sources are relatively close to the surface and the fractures occasionally penetrate water aquifers used for drinking or agricultural applications, making the fresh water effectively useless. In addition to water contamination, hydraulic fracturing has also been linked to earthquakes in Britain and the United States and has led many countries and local jurisdictions to put moratoria on further exploitation using this technique until there is proper monitoring and adequate scientific assessment to ensure public safety and environmental protection. Exploiting these non-conventional resources should be viewed as a sign of our desperation to maintain natural gas production regardless of the environmental (and ultimately social) consequences.

There is a sort of Pareto's Rule being illustrated in non-conventional gas production. Pareto noted in his economic and sociological studies that an 80/20 rule seems to dominate many situations: 20% of the population controls 80% of the wealth; 80% of the absenteeism can be attributed to 20% of employees; 80% of errors come from 20% of the politicians ... okay, it's not a perfect rule. In the case of natural gas, we can expect that 80% of the environmental damage will occur from this last 20% of non-conventional gas production. We're effectively burning down the house to keep us warm.

There is a lot of optimism, however, about the amount of natural gas that can be liberated from formations using fracturing methods of stimulation. The potential reserves of oil-

and gas-bearing shale in North America and in other regions of the world are truly staggering. The amount of oil and gas that can actually be produced from these formations is still uncertain, and many believe that current projections are overly optimistic. The abundance of shale gas reserves may push the graph a little more into the future, maybe a decade or two, but the general trends should be considered accurate.

Coal

The use of coal as a fuel is much more complex than either oil or natural gas. There are a number of grades of coal including (from highest carbon content to lowest): anthracite (86-97% carbon), bituminous (45-86% carbon), sub-bituminous (35-45% carbon), and lignite (35-45% carbon). The heating value (energy when burned) is related to the amount of carbon in the coal. The heating value for anthracite is about 30 MJ/kg, and this value drops to 19 to 30 MJ/kg for bituminous, 8 to 25 MJ/kg for sub-bituminous, and 5 to 14 MJ/kg for lignite. The lower the grade of coal, typically, the more pollutants generated per unit of energy produced, including greenhouse gas emissions. There are very large proven reserves of coal worldwide, with 155 billion tonnes of anthracite & bituminous coal and 103 billion tonnes of sub-bituminous & lignite.

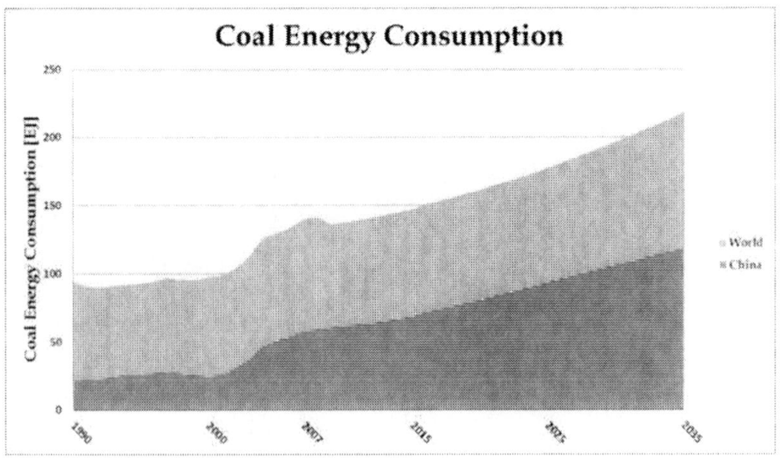

Figure 10 - Coal Energy Consumption

The worldwide consumption of coal is projected to continue to increase, with the greatest growth in China as a result of their effort to fuel rapid economic expansion. In fact, most of the worldwide increase in coal consumption can be attributed to China as they use up their significant domestic reserves, whereas coal consumption in the rest of the world has remained relatively flat ... likely an ongoing trend while oil and gas production is sustained.

Coal production, however, has a limitation that is similar to oil and gas: it is limited by the rate at which the reserves can be produced. This is related to the location, depth, and quality of the coal. The Energy Watch Group has predicted peak production of coal to occur in the first half of this century.

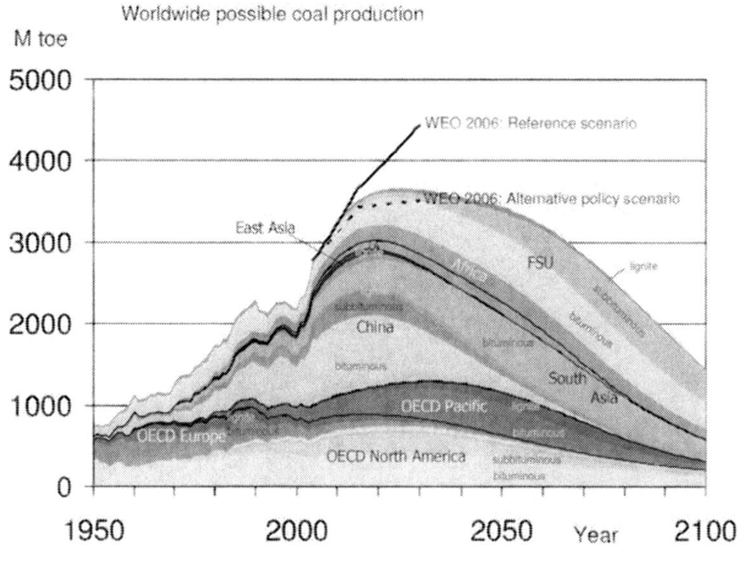

Figure 11 – Worldwide Possible Coal Production (courtesy of Energy Watch Group, www.energywatchgroup.org)

At the risk of sounding repetitive, the peaking of coal production does not mean that we have run out. There will always be enough coal for Santa to leave in your stocking each year.

We have briefly introduced the limits of exploiting oil, natural gas, and coal based on geological and technical limits. These three fossil fuels constitute 90% of our current global energy consumption, and it is upon these three fossil fuels that our current civilization is energized. Another approach to evaluating our energy future, which comes to the same conclusion, is to examine the energy it takes to get more energy.

Energy Return over Energy Invested (EROEI)

It's not dark yet, but it's getting there.

Bob Dylan

The world consumption of energy is primarily satisfied using fossil fuel sources - oil, gas and coal. As noted, these resources are not renewable in a humanly meaningful time span. They are, quite simply, a one-time inheritance of stored solar energy. The proven reserves of all three of these sources of energy are massive, but our appetite for them is equally massive. It has also been discussed that there are limits to the *rate* of production of these resources. These limits include *geological limits* as we exploit smaller and more difficult to produce sources; there are *spatial limits* due to the location of the remaining reserves relative to the desired point of use; and there are *energy/economic limits* to processing lower quality sources.

All of these limits can be related to the Energy Return over Energy Invested (EROEI) of extracting, transporting, and refining the fossil fuel. Every source of energy requires an investment of energy to take it from the earth and distribute it to the energy users in the quantity and quality that the consumer requires. Oil and natural gas, for example, are made up of hydrocarbon chains of different lengths. Each type of carbon chain has unique properties - these carbon chains include methane with one carbon and four hydrogen atoms, (CH_4), ethane (C_2H_6), propane (C_3H_8), butane (C_4H_{10}), pentane (C_5H_{12}), Hexane (C_6H_{14}) and so on. Methane is the main component of natural gas used in the home for heating; ethane is a difficult waste gas that is often burned at the refinery or used as a feedstock for ethylene plastics (for milk jugs);

propane and butane are typically sold separately as fuels; pentane and the next number of carbon chains constitute gasoline; longer carbon chains make diesel; and very long chains are used for asphalt products like roads and roofing shingles. When a well is produced, all of these carbon chains are mixed together in varying amounts along with water, and often gases like carbon dioxide and sulfur compounds. The refining process separates the different carbon chains into desirable products. The refining process also removes the undesirable components like water, carbon dioxide, and sulfur (because they produce gases when burned that create acids which can become a source of corrosion in the processes that are burning the fuel, not to mention acid rain when it is emitted into the atmosphere; and because they may be emitted in the form of highly poisonous compounds like hydrogen sulphide).

It takes energy to drill a well into an oil or gas bearing zone, and to excavate coal in a mine, in an open pit, or by blasting off the top of a mountain: the smaller the deposit, the more energy per unit of production must be invested. It takes energy to transport the product: the more remote or inaccessible the source, the more energy per unit of production it requires. And it takes energy to refine it into useful grades: the lower the quality of feedstock, the more energy per unit of production it requires. Once refined, it takes energy to transport the product to the final user, and this user will burn the product at varying efficiencies to manufacture products, make electricity, fuel cars and trucks, or to heat homes, commercial facilities, and industrial buildings - everything that we rely on to maintain our current Lifestyle. The amount of energy in the final product (gasoline, natural gas, etc.) compared to the amount of energy invested in getting the energy product to the user is called the EROEI (Energy Returned over Energy Invested).

As an example of EROEI, imagine that you live next door to your place of work, there is no dress-code, and they feed you lunch while you are at the workplace. The amount you have to invest to go to work is very little - you have a high salary-return-for-salary-invested. A year later, the company cancels the lunch program - the 'SROSI' drops a little because now you have to spend some of your salary on food for lunch. The following year, the company establishes a dress-code - you now have to invest a little more of your salary into clothing to keep your job. And the following year the company wants to expand, so they expropriate your house but give you another in the suburbs. Now you have to buy a car to get to work - another reduction to your 'SROSI'. As more and more of the employee's salary is invested in the means to maintain a salary, less is left over for personal expenditure. This is the same for energy.

The EROEI can vary widely depending on the quality of the source being extracted, the distances travelled, and the amount of processing required to convert it into a useful energy product. Because if this, it is important to realize that most published EROEI values are generalizations. In principle, however, the EROEI provides a useful indicator of the efficacy of the source of fuel. Consider, for example, the history of oil production. The first oil reserves exploited were large, accessible, light (mainly made up of smaller carbon chains), sweet (free of contaminants, like sulfur), and with geologies that allowed for high rates of production for each well drilled. Some of this oil was so easy to find, that there is a story[18] about a man named Jed, a poor mountaineer, barely kept his family fed, then one day he was shootin' for some food, and up through the ground came a bubblin' crude. This source of oil required small amounts of energy to discover, produce, transport and refine which resulted in an EROEI of around

100:1 (meaning that for every one unit of energy invested, 100 units of energy were delivered to the final user). As each oilfield was produced and began to go into decline - following a bell curve - new oil fields were developed to meet the shortfall and to match growing demand. The priority was on developing the most energy efficient (and most profitable) sources first. As a result, over the past 100 years, the EROEI for conventional oil production has dropped to less than 20:1 on average in North America. Natural gas has a similar EROEI as oil, and coal today is in the 30:1 to 60:1 range, which continues to make it very attractive for exploitation.

Unconventional sources of oil such as tar sands and shale oil require a lot of energy to extract and refine. It has been suggested that feedstock from the Alberta tar sands has an EROEI of less than 2:1, and may become even worse as the industry begins to develop bitumen deposits too deep to extract from open pits. To make the bitumen runny enough to produce (lowering the viscosity), large amounts of steam must be used to heat the bitumen and separate it from the sand that is extracted from open pits with the tarry bitumen. The steam is presently generated using enormous quantities of natural gas. The mined product has long carbon chains that must be broken into smaller chains through an upgrading process. The bitumen or the upgraded oil is mixed with a light diluent that reduces the viscosity enough to be economically pumped through pipelines to refineries across North America – for example, the Keystone XL pipeline spanning thousands of kilometers across important fresh-water aquifers and environmentally sensitive areas to Texas. These refineries convert the feedstock from the tar sands into useful fuel, like gasoline, or products for industrial processes.

Large volumes of contaminated water is created in the extraction process and must be stored in massive tailing ponds (currently exceeding 70 square kilometers in northern Alberta alone) awaiting treatment. We seem to have no problem creating the toxic tailings, even though the technologies required to treat the water remain undeveloped at a commercial (cost effective) scale. The Pareto 80/20 Rule seems to apply for the Alberta tar sands, too: 80% of the pollution and environmental impact, for 20% of the energy delivered.

The EROEI is a key to understanding why fossil fuels are so widely used compared to other sources of energy. The greater amount of energy you can sell for an initial energy investment, the greater will be the profits. When you have an EROEI of 100:1, each unit of energy consumed ('invested') can yield 100 more units of energy. As the EROEI for fossil fuels decline, more energy is required to produce the next 100 units of energy. For example an EROEI of 20:1 means that five units of energy are now used to produce 100 units of energy. As the EROEI approaches 1:1, you are using the same amount of energy to produce the product as the energy value of the product you deliver - kind of like me paying you a dollar to give me a dollar. All types of energy, however, are not equal from the perspective of the end-user. For example, it may be economically worthwhile to burn natural gas to produce oil for transportation - this is a conversion of energy from gas to liquid, which may be more valuable to the end user. This is like exchanging a dollar bill for a dollar coin that you need for a vending machine. You might even pay someone a couple of dollars for a dollar coin, if you are desperate enough. How desperate are we, I wonder?

The discussion of the EROEI of non-fossil fuel based energy will be expanded later. In the meantime, it should be noted that it is the EROEI that determines the rate of peak production. It has been argued that technological improvements in the discovery, drilling and extraction of oil and gas will result in ever greater production rates. Some say that these improvements will allow for greater exploitation of smaller discoveries. It has also been argued that coal can be converted to gas or oil through processes of gasification or liquefaction (assuming that coal reserves were infinite). These perspectives may be true in principle, but these processes require a larger initial energy investment (not to mention monetary investment in new infrastructure) and they result in ever lower EROEI values: the net energy produced (the amount delivered minus the amount used to extract, refine and transport the product) steadily declines. The decline curves in the production of fossil fuels (previously shown) become even more striking when you consider that a larger share of the energy production is being directed back into the loop for future production.

The historical EROEI for energy used by civilizations in the past has followed a bell curve similar to the one introduced by Hubbert. Early civilizations burned wood found near at hand. It is said that the Roman Empire achieved an EROEI of 20:1 at their zenith followed by a decline in both EROEI and their civilization. The application of coal in Britain at the beginning of the Industrial Revolution, which followed a crisis in 'peak wood', initiated a rapid growth in EROEI as industrial technologies evolved, and the EROEI peaked in the early to mid-twentieth century with the discovery and development of oil and natural gas. The EROEI has subsequently begun to decline. Since it takes more energy to make energy, the question is not what we *can* do (technologically); the question is if it is worthwhile to do so (from a net energy perspective). Producing low EROEI fuels is an act of desperation, and should serve as a clear warning as to the future of energy use.

To summarize, we are consuming our one-time inheritance of fossil energy at unsustainable rates, and keep in mind the magnitude of our energy consumption: it is currently an astounding 475,000,000,000,000,000,000 joules of energy each year. We are squandering this inheritance on the greatest binge party the earth has ever seen. The news of the party has spread to every corner of the world. The ever-increasing number of people arriving to enjoy this largesse is, however, exceeding our ability to provide. Instead of shutting down the party, however, we are pouring aftershave into the punch bowl. The reality of peak fossil fuel is that we will have to make do with less of it. We may be able to supplement some of this decline in the energy available by introducing more sources like nuclear, wind, solar, biomass, and geothermal. We may be able to reduce demand by voluntarily controlling our population. We may be able to improve the efficiency of energy use or invent new technologies that will extend the fossil energy production curves. Perhaps scaling back on rates consumption by shrinking our economy would help. Though it remains to be seen if we can extend our use of fossil fuels, it will be argued that there are no viable alternatives to meet an increasing appetite for energy. Furthermore, it will be shown that continuing to consume fossil fuels at current rates will have profound effects on the livability of our planet.

Minerals in Our Dirt

It's as easy as digging up some dirt and separating out all the minerals that we use in the production of our contemporary lifestyle - and we have lots of dirt, don't we? We currently mine over 30 million kilotonnes (30,000,000,000,000 kg) of minerals and metals from the earth each year to satisfy our consumption 'needs'. That's a bull elephant of dirt mined each year for each person on earth. From this, over 200 kg of metals are excavated per person each year on a global average.

This rate of extraction, of course, has a significant impact on the natural environment because so much earth must be excavated from the ground to provide the minerals and metals needed for an industrial economy. Not unlike fossil fuels, the rate of extraction depends on the quality of the deposits being mined and the accessibility of the location. Ore grade for many metals has been steadily declining in the past number of decades, since the most productive deposits have already been exploited. This has forced mining companies to move on to lower quality or more difficult to access sites. Mining lower quality ores requires significantly more energy because more earth must be excavated to get the same amount of ore for processing. The decline in cheap and accessible fossil fuel energy will result in an even greater decline in the rate of mineral extraction. It should be noted that we are not saying that there are no more minerals in the earth (like we are not saying there are no more fossil fuels) - we are talking about the rate of extraction, and the energy consumed to do so: these are the limits, and some metals have already peaked in the rate of production.

In addition to the reduction of mineral extraction per energy invested, the estimated availabilities of metals (based on the declining rate of extraction at current rates of consumption) are limited. Iron is, perhaps, one of the most important and widely used mineral on the planet. It is also relatively abundant with an estimated 50 to 100 years of availability, though, as we have learned, the size of the reserves is not as important as meeting the rate of demand. World iron production peaked in the early 2000s[19], putting much more pressure on recycling to meet demand. Based on Wouters & Bol's study[20] on material scarcity, silver has roughly 10 to 30 years of availability; gold has 10 to 45 years; cadmium has 20 to 50 years; copper has 25 to 60 years; tin has 20 to 50 years; and so on - you get the point. These are the minerals that we use to make all of our stuff. Some materials are replaceable by other materials, but all the materials are not replaceable by nothing (which is all that's going to be left).

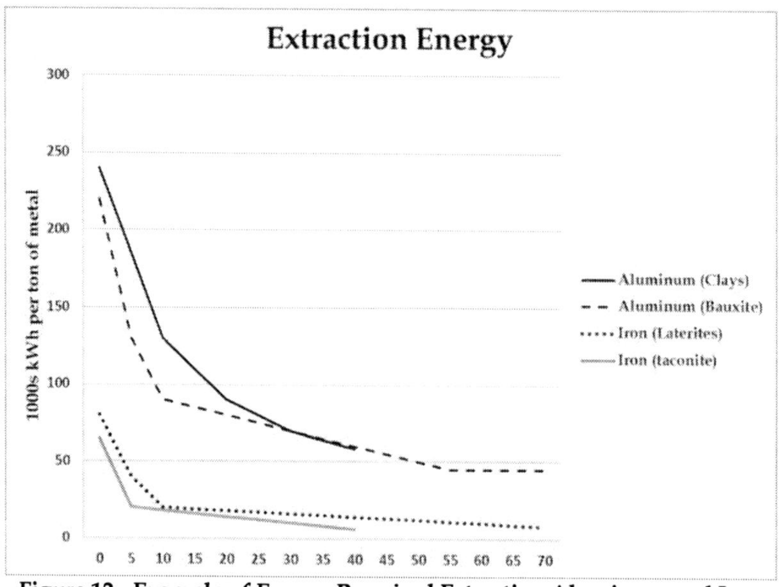

Figure 12 - Example of Energy Required Extracting Aluminum and Iron

It has been suggested that improvements in mining techniques and technologies will increase recoverable reserves; that conservation and productivity improvement will reduce demand; and that better recycling will capture more in-cycle metals and reduce the demand on virgin sources. Sure. But from where is the energy going to come to keep producing these minerals and metals? As the ore grade diminishes, the energy required to extract the ore increases exponentially, as shown.[21] There is a real limit for mining minerals and metals in a world of declining rates of fossil fuel production and EROEI.

And consider more closely those timelines for extractable resources - three, four, five decades! When we speak about extending these resources, are we talking about an additional decade or two? That sort of discussion reminds me of my grandparents arguing if they visited the Grand Canyon on a Tuesday in 1957 or a Thursday in 1958.

The point is that we are consuming some very important metals and minerals at unsustainable rates. In fact, all mining is unsustainable but it might be justifiable if we were to keep as much of the material in circulation before it is finally exhausted and sent to the landfill (remember the law of entropy). Most of the reserves for these important metals and minerals are being found in more remote regions and at lower concentrations (ore grades), requiring considerably more energy to produce. And, keep in mind; there will be less energy available to use in the future. In other words, you can't rely on getting scarce and remote metals and minerals when there is no energy available to do so.

The discussion so far has focused on the energy and resources that we use in our consumption-based lifestyles. The numbers lead to a grim forecast for the future - but it mainly only effects our lifestyle. Less energy and fewer metal and mineral resources will profoundly affect our way of life, but not necessarily our lives. What follows is a discussion of our impact on the earth systems that sustain our lives - and, well, the story doesn't get any better.

The Great Big Human Footprint

A group of leading scientists, including a number of Nobel laureates, have identified nine "planetary boundaries" that are crucial to sustainable existence on the planet[22]. Foster, Clark & York present them as follows:

> Climate change is only one of these, and the others are ocean acidification, stratospheric ozone depletion, the nitrogen and phosphorous cycles, global fresh water use, change in land use, biodiversity loss, atmospheric aerosol loading, and chemical pollution. ... Three of the boundaries - those for climate change, ocean acidification, and stratospheric ozone depletion - can be regarded as tipping points, which at a certain level lead to vast qualitative changes in the earth system that would threaten to destabilize the planet, causing it to depart from the "boundaries for a healthy planet." The boundaries for the other four processes - the nitrogen and phosphorous cycles, global fresh water use, change in land use, and biodiversity loss - are better viewed as signifying the onset of irreversible environmental degradation.[23]

Beginning with climate change: The preindustrial concentration of carbon dioxide in the atmosphere was 280 parts per million (ppm); its boundary, as proposed by James Hansen, is 350 ppm; and the concentration is currently 390 ppm. Without immediate and significant efforts in reducing emissions, the tipping point for climate change is imminent.

The second boundary, ocean acidification, is regulated by the concentration of calcium carbonate (from limestone in surface runoff) in seawater. Under normal conditions, as carbon dioxide is absorbed by the ocean to form carbonic acid, the resulting acidity is neutralized by the calcium carbonate. Think of taking an antacid after overindulging in your favourite Mexican dish. This 'buffering' effect will occur as long as the calcium carbonate lasts. The preindustrial value for aragonite (a form of calcium carbonate) was 3.44 and the level is currently 2.9. The boundary limit is estimated to be 2.7, after which the health of shell-forming organisms becomes threatened, not to mention the species that eat these organisms. The third boundary that might act as a tipping point, stratospheric ozone depletion, remains a serious concern though this threat has been identified and generally accepted, with worldwide efforts to stabilize and reduce the most harmful emissions.

With respect to the nitrogen and phosphorous cycles, the boundary relates to the amount of nitrogen removed from the atmosphere for agricultural use. Before the discovery of the Haber-Bosch process in the early 1900s, agriculture relied on natural sources of fertilizers, particularly guano. There is an interesting history around the guano trade including slavery, the making and breaking of countries in South America, and even war, so desperate were countries to sustain agricultural productivity for their populations. Since the early twentieth century, however, nitrogen has been simply removed from the atmosphere and fixed in the soil as fertilizers to enhance plant growth. Only a small part of this nitrogen is used by plants, the rest pollutes water systems or is emitted in more reactive forms into the atmosphere (as a potent greenhouse gas). This process ultimately disrupts the nitrogen cycle of the earth. The boundary proposed to avoid irreversible degradation of earth systems, is 35 million tons per year and the current rate exceeds 120 million tons per year.

The preindustrial quantity of phosphorous flowing into rivers, lakes and oceans was approximately 1 million tons each year, which is well below the proposed boundary of 11 million tons (based on the assumption that ocean anoxic events begin at ten times the background rate). The current rate of phosphorous flowing into our water systems approaches this boundary at 8.5 to 9.5 million tons each year. The good news is that world rock phosphate production peaked in 1989 and is now (and forever) in steady decline[24] - so, we will not likely exceed the estimated boundary. The bad news, of course, is that phosphorous is an essential component of agricultural yields to feed a growing global population – there is no substitute for phosphorous in agriculture. More on that, later.

Freshwater use was 415 km^3 (cubic kilometers) in preindustrial times and our current consumption 2,600 km^3. The estimated threshold for freshwater abstraction before irreversible degradation and the collapse of aquatic and riparian ecosystems is estimated to be 4,000 km^3. Concerns about water availability are difficult to appreciate when there appears to be abundance. Barlow & Clarke describe the world's available water as follows:

> The total amount of water on earth is approximately 1.4 billion cubic kilometers ... The amount of *fresh* water on earth, however, is approximately 36 million cubic kilometers (about 8.6 million cubic miles), a mere 2.6 percent of the total. Of this, only 11 million cubic kilometers (about 2.6 million cubic miles), or 0.77 percent, counts as part of the water cycle in that it circulates comparatively quickly. However, fresh water is renewable only by rainfall. So in the end, humans can rely only on the 34,000 cubic kilometers (about 8,000 cubic miles) of rain that annually form the "runoff" that

> goes back to the oceans via rivers and groundwater. This is the only water considered "available" for human consumption because it can be harvested without depleting finite water sources.[25]

Certainly, the 34,000 cubic kilometers that is replenished by the hydrologic cycle is a lot of fresh water but, unfortunately, much of this water falls where the people who need it are not. The amount of water flowing in rivers worldwide is declining at important times of the year due to the loss of mountain glaciers (a result of global warming) or it flows when it is least desirable (early melt of the snowpack in spring and rapid runoff due to deforestation in the headwaters). Furthermore, much of this water must be left in lakes and rivers to support aquatic and riparian life. Barlow concludes:

> the verdict is in and irrefutable: the world is facing a water crisis due to pollution, climate change and a surging population growth of such magnitude that close to two billion people now live in water-stressed regions of the planet. Further, unless we change our ways, by the year 2025, two-thirds of the world's population will face water scarcity. The global population tripled in the twentieth century, but water consumption went up sevenfold. By 2050, after we add another three billion to the population, humans will need an 80 percent increase in water supplies just to feed ourselves. No one knows where this water is going to come from.[26]

What is worse than the unhappy combination of increasing water demand with unpredictable supply is the fact that we have been polluting much of the freshwater remaining available to us. Consider the following summary from Bob Willard:

> In Eastern Europe, there's hardly a river, stream, or brook that isn't contaminated … In China, 80 percent of the country's 50,000 kilometers of major rivers are so degraded they no longer support any fish. … There are more than 8,000 kilometers of shoreline on the US side of the Great Lakes, but only 3 percent are fit for swimming, for supplying drinking water, or for supporting aquatic life. … Each year 50 - 100 million tonnes of hazardous waste is generated in the watershed for the lakes, 25 million tons of pesticides alone …[27]

Maude Barlow similarly describes the effluent from water being returned to the ecosystem after what is considered 'treatment':

> Pesticides, industrial wastes, arsenic, and metals all showed up in the "treated" water flowing into the St. Lawrence River. The Quebec study observed that 'in all, more than 85 percent of the sewage samples from all sources contained the following: Ammonia, phosphorus, aluminum, arsenic, barium, mercury, PCBs, chlorinated dioxins, and furans, surfactants (cleaning chemicals), polyaromatic hydrocarbons (PSHs), and other organic and inorganic wastes.'[28]

In addition to consuming and polluting the fresh surface water that we need, we have been over-consuming groundwater from aquifers which are not replenished in meaningful time periods. Pearce says: "We are already living on borrowed time by mining the aquifers - dipping into the slow, often largely unrenewable water cycle to top up the fast, renewable cycle. ... Up to a billion people are today eating food grown using underground water that is not being replaced."[29]

In other words, we are using the world's fresh water at unsustainable rates, at rates that diminish the ability of the ecosystem to flourish. We are using water from 'fossil' aquifers to supplement our consumption, withdrawing fresh water from the one-time inheritance of underground water. And what we don't use we destroy by using lakes, rivers and oceans to dilute our chemical pollution. In some rivers, we are abstracting water at such a high rate that they actually stop flowing: "The following rivers no longer reach the oceans in the dry season: the Colorado, Yellow, Ganges, and Nile, among others."[30] In other rivers, the flow remaining when the river meets the ocean is a chemical stew not fit to support aquatic or riparian life. Good grief.

The change in land use is another planetary boundary. The percentage of arable land is approximately 15% of the ice-free surface on earth. The amount used for agriculture has risen from very low levels in preindustrial times to as much as 80% of the total available today. The conversion of land to agriculture or pasture has meant (and continues to mean) vast amounts of deforestation, and a huge depreciation of organics held in the soil. All of this stored carbon has entered the atmosphere as greenhouse gases in the form of carbon dioxide and methane. The limit, however, is clear - once all of the arable land is converted to agricultural use to feed our growing global population - it's done: 100%. At this limit, it will be necessary to increase and sustain the productivity of food crops, and to avoid the further erosion and desertification of the productive land that we will rely on to sustain us - and, as will be argued later, these things we either cannot do, or will not do.

Biodiversity loss is measured by the number of species lost per million known species each year. The preindustrial annual rate of extinction, referred to as the 'background rate', was 0.1 - 1 per million and the current rate is greater than 100 per million (100 - 1000 times the preindustrial background rate). There is probably no appropriate boundary beyond the background rate of extinction.

> The MA [Millennium Ecosystem Assessment] explains in measured, scientifically-correct statements that human beings are having a devastating impact on virtually all ecosystems. It points out that if one applies the IUCN-World Conservation Union criteria for threats of extinction, currently 12% of bird species, 23% of mammals, 25% of conifers, at least 32% of amphibians, and 52% of cycads (evergreen palm-like plants) are threatened with extinction. In the last few decades of the 20th Century alone human activities destroyed at least 35% of Earth's mangrove areas (vital breeding grounds for many marine fish), 20% of the world's coral reefs and degraded a further 20% of coral reefs.[31]

Finally, there are no adequate physical measures for atmospheric aerosol loading and chemical pollution at this time. That does not mean that there are no impacts of these pollutants on the health of the ecosystem, nor does it mean that there are no planetary boundaries after which the ecosystem fails. Although there are hundreds of thousands of synthetic chemicals, there are roughly 62,000 that constitute the bulk of chemicals used in the United States. Of these, 3,000 chemicals comprise over 99% of the 15 trillion pounds manufactured or imported each year[32]. There is little known about the interactions of chemicals, the impact of exposure in the workplace and home, the effect of bioaccumulation in the environment, and the impact on early-life exposures - but we use them anyway. The cost of health is not known, but the United States spends over $1 billion each year to manage superfund sites, of which many involve the clean-up of toxic chemicals abandoned by industry.

Those are the planetary boundaries, beyond which we risk upsetting the natural systems upon which our civilization relies.

> Nature is a ready storehouse of the raw materials of civilization - food, fiber, fuel, minerals - and the collective annual value of these goods is in the trillions. But the global ecosystem also provides many services that are the indispensable substrate of economies, including air and water purification, mitigation of droughts and floods, soil generation and soil fertility renewal, waste detoxification and breakdown, pollination, seed dispersal, nutrient cycling and movement, pest control, biodiversity maintenance, shoreline erosion protection, protection from solar ultraviolet rays, partial climate stabilization, and moderation of weather extremes.[33]

Another point is that any single one of these boundaries could upset the dynamic of our whole system and challenge our very existence: "Liebig's Law named after the 19th century German soil scientist Justus von Liebig, is sometimes called the Law of the Minimum. It tells us that the carrying capacity for any given species is set by the necessity in the least supply. Every species has a list of requirements for survival - water, temperature range, degree of salinity of water, degree of acidity or alkalinity of soil, food of a certain nature, so many hours of sunlight, and so on. Liebig's Law tells us that when if all other factors are optimal, the lack of one necessity can undermine an organism's ability to survive."[34] Heinberg goes on to observe the hazard of relying on our ability to manage planetary boundaries and the reality of Liebig's Law should dispel the vain notion of creating a fully artificial environment.

The biosphere is a collection of complex systems that must remain in balance to sustain life as we know it. The planetary boundaries outlined above involve these systems - the carbon cycle, nitrogen cycle, hydrologic cycle, and phosphorous cycle - and the planetary boundaries relate to the substances we require in order to continue to support the world's population of humans (and, oh yeah, those other species, too) - fresh water, clean air, fertile soil. All of these boundaries are important: we could fix eight of the nine planetary boundaries that we are approaching or have already exceeded, and still fail because of our collective inability to address the ninth. This ninth boundary, the one that we cannot or will not fix, will likely be climate change caused by global warming. As such, this topic will be expanded next, and the reasons why we will stumble over this planetary boundary will follow.

Already Extinct

"If today is a typical day on the planet Earth, … we will lose 116 square miles of rain forest, or about an acre a second. We will lose another 72 square miles to encroaching deserts, the results of human mismanagement and overpopulation. We will lose 40 to 250 species, and no one knows whether the number is 40 or 250. Today the human population will increase by 250,000. And today we will add 2,700 tons of chlorofluorocarbons and 15 million tons of carbon dioxide to the atmosphere. Tonight the Earth will be a little hotter, its waters more acidic, and the fabric of life more threadbare."[35]

David Orr

Climate Change

> We are crossing natural thresholds that we cannot see
> and violating deadlines that we do not recognize.
> Nature is the time keeper, but we cannot see the clock.
>
> Lester Brown in *Plan B 3.0*

Climate change, like peak oil, has provoked considerable resistance and a concerted effort in denial. The denial industry is actually a small group of individuals (including some scientists from unrelated disciplines, or economists and political scientists who think they are scientists, and no one has the heart to tell them they are not) who have been amply funded by industries threatened by a shift away from fossil fuel consumption. Some of these deniers have impressive pedigrees of noble causes going back to the denial of ozone depletion and rejecting the risks of smoking tobacco.[36] Sadly, this well-funded attempt to create doubt has delayed what might have been a timely, and more cost-effective, intervention towards a more sustainable relationship with the earth that is a source of resources and a sink for our wastes. More sadly, polls suggest that the efforts in manufacturing doubt have been quite effective in North America, and will very likely defer any meaningful response in the future. The impotence of the climate change discussion and international efforts to curb global warming has been made manifestly clear in recent years. Pandora out of curiosity opened her jar and released all the evils of mankind before closing the jar again. The only thing left was hope. It seems that the international community of politicians and orthodox economists have found Pandora's jar and have sealed it so well that hope will never be released.

The actual science of the greenhouse effect has been with us for well over a hundred years. Quite simply, some gases absorb heat and the resulting greenhouse effect warms the earth - what some call global warming. Radiation from the sun enters the earth's atmosphere where much of it is reflected back into space or re-radiated as infrared radiation. The radiation that is not reflected back into space is absorbed by the surface of the planet, which heats up and reradiates energy in the infrared portion of the spectrum. Molecules in the atmosphere (primarily carbon dioxide, methane and water) absorb some of this radiation and trap the heat in the atmosphere - like a greenhouse. This is a good thing: without the greenhouse effect the average temperature of the earth would be -18°C, which would not be conducive to the development of intelligent life, like dolphins. Clearly the greenhouse effect is real and has a significant impact on global warming. This is accepted science, even by all but the most obtuse of deniers. It is difficult to not accept the next logical conclusion: that the addition of greenhouse gases to the atmosphere from human activities will further increase average global temperatures.

It should be noted that an increase in average global temperature is not the same as the normal daily fluctuations of local temperature. As I write this, the temperature is rising from a morning low of -25°C to a balmy -1°C. It will cool off again tonight to a forecasted low of -12°C. These daily fluctuations make the slow change in the average global temperature appear nonsensical, but it is this slow change that will have a profound effect on the biosphere and how our ecosystems function.

The addition of greenhouse gases to the atmosphere (from natural sources, from deforestation or cultivation, or from the burning of fossil fuels) at rates above the ability of earth systems to absorb will increase the average temperature of the earth. Industrial civilizations are currently adding considerably more carbon dioxide, methane and other complex chemicals to the atmosphere each year than nature can absorb (remember 475 EJ of energy used each year, primarily from fossil fuel burning). The addition of every 2.12 Gigatonnes (billion metric tonnes) of carbon is estimated to increase the carbon dioxide concentration in the atmosphere by about 1 ppm (part per million). Though the carbon cycle is dynamic, with some systems absorbing carbon and other systems emitting carbon, the net increase in carbon to the atmosphere is currently about 4.5 Gigatonnes each year. This equates to an increase of 2 ppm each year, a human climate forcing which James Hansen says is four orders of magnitude - ten thousand times - more powerful than normal.

The ability of earth systems to absorb carbon (even if we were able to eliminate the further burning of carbon fossil fuels) depends on a number of factors including reforestation, sequestration in soils, and the health (acidity) of the oceans. As emissions of greenhouse gases to the atmosphere increase, the gas concentrations increase because they cannot be absorbed by earth systems. This is exacerbated by the fact that the absorption of greenhouse gases by earth systems is diminished when the natural cycles are damaged by human impact.

The warming effect of additional emissions of greenhouse gases is also not linear, which means that each 2.12 gigatones of carbon added to the atmosphere cannot be related directly to increases in the average global temperature. For example, the gases present in the atmosphere may interact with earth systems to create an amplifying effect. One example is the addition of carbon dioxide to the atmosphere which, by itself, adds to the warming effect, but it also increases the amount of water vapor in the atmosphere which significantly adds to the forcing.

Furthermore, the warming effect is not linear with emissions or the concentration in the atmosphere[37] - this means that doubling the concentration doesn't double the amount of heat being contained in the atmosphere. This is because at a certain concentration of greenhouse gas all of the available energy in a certain part of the infrared band will have been absorbed. At this point, further increases in those emissions will not increase the amount of heat absorbed (there is no radiation in that portion of the spectrum left to absorb). For example, if a given concentration of a greenhouse gas already absorbs 80% of a certain wavelength in the spectrum, doubling the concentration does not double the amount of the wavelength absorbed to 160% - the maximum is 100%, and the amount of greenhouse gas above this 'saturation' concentration has no direct effect in warming (though the carbon dioxide is still absorbed by oceans, intensifying acidification and reducing its ability to absorb carbon while driving a good proportion of ocean species to extinction).

What makes this even more complex: in addition to 'saturation' there is 'spectral overlap' which describes the phenomenon that the wavelength absorbed by different greenhouse gases overlap. For example, imagine a dozen machines shooting tennis balls onto a court, one person (let's call her 'carbon dioxide molecule') can hit back only the balls in her vicinity, and at the rate she can swing the racket. If you keep adding people, more balls are going to be hit back. Presumably, with enough people, all of the balls will be returned from that side of the court ('saturation'). Now imagine adding different gases - I mean people - some tall ones at the back of the court (let's call them 'methane molecules'), and a few to stand at the net to catch the short lobs (they are 'nitrous oxide molecules'). Adding a few taller people to hit the high ones, and putting a few people at the net to hit the short lobs may improve the returning of balls, but the other people will have fewer to hit ('spectral overlap').

As an example, research on greenhouse gases show that removing all of the carbon dioxide from an atmosphere will reduce infrared absorption by only 9%. Removing all other gases except carbon dioxide will reduce infrared absorption by 74% (leaving 24% absorption). So, does carbon dioxide absorb 9% or 24%? This depends on what other gases are available. It should also be noted that the presence of one gas, like carbon dioxide, may increase the concentration of another gas, like water vapour, which has a much greater ability to absorb infrared energy. Though complex, the science of greenhouse gases and their impact in the atmosphere is quite well established, despite some efforts to confuse the issue.

That the impact of emissions in the atmosphere is complex does not suggest that warming is not occurring or that the science is not reasonably accurate. That's like saying heart surgery is complex therefore no surgeons can perform the operation, and even if they do, it's so unpredictable that you are likely to die. These complexities simply suggest a range of scenarios of possible rates of warming (of which, it should be noted, all threaten to upset the steady-state balance that has been enjoyed throughout human civilization). Uncertainty does not mean doubt - if I am walking to work, I will be there in 15 minutes with an uncertainty of a plus or minus a few minutes with an exception once a month or so if I meet a friend along the way. It does not mean that I won't make it to work. All research has uncertainty, because it is statistical and there is variance in everything that is measured.

To repeat, scientific uncertainty does not mean doubt.

The increase of carbon dioxide concentrations in the atmosphere from emissions of burning fossil fuel is exacerbated by clear cutting forests and converting natural grasslands to agriculture, because both forests and natural soils absorb and hold (sequester) huge amounts of carbon. As the oceans become more acidic, the ability to absorb carbon dioxide decreases. Worse, it is believed that when a certain level of acidity has been reached, the oceans may become net emitters of carbon dioxide, a positive feedback with dire consequences because the ocean has traditionally been a major sink for carbon dioxide. Other positive feedbacks that could lead to runaway average global warming and climate change include the loss of ice due to the melting of glaciers and the ice caps, which reduces the reflectivity (albedo) of the earth. As the ice disappears, less radiation is reflected back into space and more is absorbed by the recently exposed water. Warming will lead to the melting of permafrost, large frozen bogs in the arctic, releasing unimaginable amounts of methane, a greenhouse gas that is much more potent than carbon dioxide in absorbing heat energy. These thresholds, or tipping points, will effectively remove our ability to control the planetary balance within the limits that life has previously thrived. Up to now, our foot has been controlling the accelerator, but at a certain point the pedal will get stuck on the floorboard and we will no longer be in control of the car - and the engine will not turn off when we turn the key (and, yes, the doors are locked, too).

With this general understanding of the science of climate change and the complexities of atmospheric science and feedback mechanisms, a brief discussion of the projections is in order. The Intergovernmental Panel on Climate Change (IPCC) is the *de facto* clearinghouse for information on the effects of our continued emission of greenhouse gases into the atmosphere.

Climate change has become a catch-all phrase that includes the effects of increasing the average temperature of oceans and the planet's surface which, in turn, intensifies weather patterns including drought, heat waves, rain and snow events, and the frequency and intensity of hurricanes and tornados. The rising average global temperature has been linked to species migration (as they try to maintain a familiar climate), rising rates of extinction, disease and pest vectors like the pine beetle infestation in western North America, and rising sea levels (due to expansion caused by warming and the melting of glaciers and polar ice worldwide). In other words, it's time to sell your beach-front property in Tuvalu.

The oceans absorb less carbon dioxide as they warm, and this is significant as they have absorbed a majority of the carbon dioxide emissions from the past few centuries of industrial capitalism. When water absorbs carbon dioxide it forms carbonic acid and lowers the pH (making it more acidic). As discussed earlier, ocean acidification is buffered by absorbing calcium carbonate which originated from water flowing over limestone through the millennia. Calcium carbonate, however, is also required by shell-forming creatures and coral development, and the competition with acidification threatens their existence. These creatures play a vital role in the food chain, and it is said that a quarter of all marine species rely on coral for habitat during their life-cycle[38]. The IPCC suggests that half of marine species are at risk of extinction in the next decades - but who likes fish, anyway.

The IPCC argues that carbon dioxide levels must be reduced to less than 350 ppm immediately to avoid the even greater dangers of runaway climate change - in which average global temperatures will increase to a new equilibrium of as much as 8°C above pre-industrial levels despite human efforts to mitigate this change. It has been said by some 'climate contrarians' that 8°C is the difference between living in Norway or in Greece, so what's the big deal? It goes to show how anthropocentric our views are. This new equilibrium, and the rate of change to this new equilibrium, will exceed the ability for most species of plants and animals to adapt, leading to massive extinction including, perhaps, the species that originated the unbalance - but who likes humans, anyway.

IPCC results show six scenarios, depending when emissions are stabilized and the rate of decrease thereafter. The worst case (business as usual) shows a stabilization of GHG emissions by mid-century. Average global temperatures are anticipated to be between 4 and 8°C higher than pre-industrial levels by the end of the 21st century. The best case is stabilizing emissions by 2015, with an aggressive reduction of 80% by mid-century and carbon sequestration by the end of the century. Average global temperatures in this case are anticipated to be between 2 and 3°C higher than pre-industrial temperatures. Some leading scientists believe that even this best-case scenario may lead to catastrophic changes - this means that we have already committed to travelling through territories uncharted by human civilization.

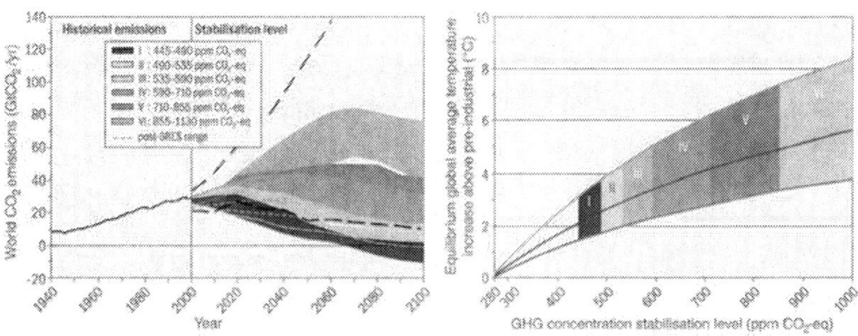

Figure 13 - Global Average Temperature Increase (courtesy of the *IPCC Fourth Assessment Report: Climate Change* 2007, Figure 5.1)

It should be noted that the IPCC predictions likely underestimate of what may transpire. The reasons for this are threefold: "governmental involvement does imply an inbuilt conservatism in the 'policy-relevant' conclusions of the IPCC, for governments do not want reports that force them into major unanticipated expenditures. Another reason for the conservatism is the natural dynamics of a consensus-based process, in which the scientists on each committee are much likelier to reach consensus at the lowest common denominator of a range of estimates about the severity of the problem than at the high end of the range. A third reason is the fact that the IPCC (which does not do original research) imposes a cut-off date after which newly published scientific studies cannot be included in the material considered by the various working groups."[39]

The coming equilibrium (whatever that may be), and the rate of change toward this new equilibrium, will exceed the ability for most species of plants and animals to adapt, leading to massive rates of extinction. Before this happens, however, we can expect increasing tensions over the declining availability of resources; disputes over clean water and arable land; and stresses caused by the migration of desperate people to other regions. This can be reframed as a fight by rich nations to preserve what they have against poor nations who will suffer from what we have caused. As Gwynne Dyer has said, "The military profession, especially the long-established great powers, is deeply pessimistic about the likelihood that people and countries will behave well under stress."[40]

To achieve the required cuts in emissions,

> it will be necessary to take action across the board and not in just two or three sectors such as power and transport. For the world as a whole, energy emissions represent around two-thirds of the total, nonenergy around one-third. Land-use change, mainly deforestation and degradation of forests, accounts for nearly 20 percent of the total. Given that the world economy is likely to be perhaps three times bigger in mid-century than it is now, absolute cuts of around 50 percent would require cuts of 80 to 85 percent in emissions per unit of output. Further, since emissions from some sectors (in particular agriculture) will be difficult to cut back to anything like this extent, and since richer countries should make much bigger proportional reductions than poor countries, richer countries will need to have close-to-zero emissions in power (electricity) and transport by 2050.[41]

This is an important point. Fossil fuel burning only contributes about half of the global total of carbon emissions. Of the other sources, roughly 20% is related to deforestation, 10% is from nitrous oxides (from fertilizers, combustion, and clearing land), and about 15% is methane emissions from livestock, sewage, and fossil energy production. So, even if one wants to reduce total carbon emissions by 80 to 85%, it is not likely to come from food production without a serious decrease in population (though some of this could be mitigated with a shift in consumption from meat to grains). The conclusion, however, is that the emission reductions will have to come from eliminating the further burning of fossil fuels for our energy needs - all of it - and the effective capture and sequestration of carbon from the atmosphere.

We'll discuss later why this isn't going to happen. By way of a summary of "*the problems that we don't understand*": non-renewable resources are non-renewable - that means finite - and a limit has a limit - that means, well, there are limits. The fossil fuels, metals, and minerals that we rely on for our very existence are limited. Worse, the *rate* of production is limited and typically follows a bell curve of steady increase to a plateau which is followed by a steady and inexorable decline. And even worse, the energy required to extract these metals, minerals, and fossil fuels increase drastically as they become more scarce or difficult to refine (and, to be clear, energy is not becoming more plentiful). Fresh water is limited to the rate of replenishment by the hydraulic cycle, and to the extent that we can rely on predictable weather patterns to provide water for agriculture and industrial demand. The ability of our water systems to absorb the pollutants that are diluted in rivers, lakes and oceans before affecting ecosystem health is limited. The ability of oceans to absorb carbon dioxide without becoming too acidic for the existence shell-forming species that represent the foundation of the food chain (including ours) is limited. The ability of the atmosphere to absorb pollutants and greenhouse gases without destroying the ozone layer (that protects us from radiation) and absorbing too much energy (that warms the planet) is limited.

The big thing that we just don't seem to understand is that we have reached those limits. But knowing isn't doing and, it will be argued, we've waited too long to do anything anyway.

… we couldn't do anything about it if we did …

We're Not As Smart As We Think We Are

Surely, now that we are aware of the problems we will do something to fix them.

Did you ever hear this one? "We proved that we could mobilize our resources during double-ya-double-ya-two," ... a knowing pause, a wink, and a confident grasping gesture of the cajones ... "and we can do it again when necessary." (If you haven't heard this one, it's the ol' war-time-mobilization-deferral-tactic with a hint of patriotic muzzling). The sad and simple reality of the statement is that we couldn't.

Fresh water? We'll make it from sea water. Arable land? We'll cut down more forests. Biodiversity loss? We'll save species' DNA and clone them later. Food security? We'll eat less meat. Global warming? We'll erect a space umbrella. Metal and mineral scarcity? We'll recycle. Energy challenges? We'll use renewable sources. Yeah, right.

To make fresh water from the sea requires energy; cutting down forests exacerbates global warming and biodiversity loss; and recycling and renewable energy requires energy up-front, and the creation of a currently non-existent infrastructure. This is simplistic, but the point is that each of these problems does not exist in isolation. You simply can't increase energy consumption to address the impacts of population, lifestyle, and climate change without exacerbating impacts of climate change. Since energy is the keystone of the discussion, let's begin there.

Alter Idem Energy?

When blindness was a virtue.

Bob Dylan

As noted earlier, we presently consume 475 EJ of energy each year, most of which comes from non-renewable fossil energy sources. Fossil energy is limited, and that rate oil production has peaked, with natural gas and coal approaching their zenith. Furthermore, according to the IPCC, we must reduce our fossil fuel consumption to roughly zero by 2050 to avoid the worst impacts of climate change. If we are facing the decline of fossil fuel availability, and if we shouldn't burn the fossil fuels even if they were available, why don't we convert to alternative energy: photovoltaic, solar thermal, wind, tidal, geothermal, and biofuels? Couldn't we switch to renewable energy sources and maintain business-as-usual? The short answer is that each of these sources has potential. At least they would have had potential if they had been embraced sooner in our industrial civilization. Each, however, is limited by the reality of net energy (EROEI) and some other consequences of technology production (what we will call embodied pollution) - the Cadmean victory of technology over nature. Let's consider each form of renewable energy in turn.

Photovoltaic (PV) Technologies

Photovoltaic (PV) means *light-electricity*. A PV array is designed to convert solar energy into electricity. The basic unit of a PV panel is the PV cell, which consists of layers of a semiconducting material such as silicon. When light is absorbed by the semi-conducting material, electrons are mobilized and directed to a circuit to make an electric current. Groups of PV cells are joined together to produce a PV module, which can be linked to other modules to create a PV array. The PV array can be housed in a structure to make a PV panel that can be mounted on a building to generate electricity.

A PV cell has two semiconductor layers: a positively charged p-type and a negatively charged n-type. The n-type semiconductor is impregnated (doped) with phosphorous atoms with free electrons, and the p-type is impregnated with boron atoms which are lacking in electrons. When light strikes the cell, the electrons in the n-type semiconductor layer are energized and they flow towards the p-type semiconductor. The electrons are caught at the junction between the n- and p-semiconductors and connected to an electric circuit. Wasn't that interesting?

There are many different types of photovoltaic production processes, including those that create the more common crystalline silicon and the evolving thin-film materials (which are cheaper to manufacture and use less polysilicon, but produce electricity less efficiently). There are many other 'breakthrough' technologies being touted by those seeking venture capital, but they are not presently contributing to the worldwide growth of PV production. To be clear, I'm not saying that these technologies will never contribute to our electricity generation potential, or that there won't be superior technologies developed in the future, that is, if we could wait a few more decades. The argument is that we don't have a few more decades to begin the process of reducing our fossil fuel use. This discussion will deal with what we can do right now with the technologies that are commercially available.

The growth of installed photovoltaic power capacity has been exponential through the first decade of the millennium. About 7300 MW was installed in 2009, adding to the existing 14,000 MW already in existence[42]. Of this capacity, over half has been installed in Germany (and almost three-quarters in Europe). The prices of PV modules are sensitive to the supply of polysilicon, as well as improvements in manufacturing efficiency. Prices are also sensitive to subsidies, and PV production has been highly subsidized to date. Warren Buffet uses the example of a housing bubble when discussing subsidies - as long as neighbours sell their houses to neighbours the prices of houses can increase endlessly. The price of homes has been set in an artificial market. As soon as someone sells their home to a person from beyond the neighbourhood, the 'real' price becomes evident. What is the real price of polysilicon? The answer is that we don't really know yet. It is expected, however, that PV prices will continue to decline as production increases (supply and demand), and it is speculated that the solar capacity will quadruple by 2020 (to about 80,000 MW), especially if China, India and Japan meet their solar targets.

What does this really mean in the global efforts being deployed to meet our future energy needs? The current installed capacity, with optimum performance, would produce roughly 21,500 GWh of electricity (at 1020 kWh/y per kW of installed capacity) - the world uses about 20,000,000 GWh of electricity each year. Doing the math, 0.1% of the global electricity demand is currently provided by PV. Now let's say we achieve the optimistic forecast (based on sustained government subsidies) of 80,000 MW of capacity by 2020 when electricity consumption is expected to reach 25,000,000 GWh - that would be a little more than a 0.3% contribution to global electricity demand. Using an even wilder prospect of a PV production growth rate of 20% each year sustained over the next decade, which would result in 250,000 MW of installed capacity, meeting 1% of our global consumption of electricity. That leaves a big gap in what we can produce and what we want to consume. At best, solar PV will be a nice technology for those rich enough to buy it (and, yes, we are talking about the same 'rich' who contributed most to the problems that have lead us to need to reduce fossil energy consumption in the first place).

It takes energy to make photovoltaic panels, and this energy is largely provided by fossil fuels (what else is there?). Studies on the energy-return-on-energy-invested (EROEI) for photovoltaic vary widely between technologies and manufacturers, but there is some consensus on the range of values. A recent study[43] determined that 2600 to 8600 MJ of primary energy is used for each square meter of photovoltaic array using monocrystalline modules; 2000 to 5600 MJ/m^2 for polycrystalline modules; and 775 to 1800 MJ/m^2 for thin film technologies. The conversion efficiency of each of these technologies declines in order of presentation, so the rate of 'energy-pay-back' is lower for each technology, respectively. Based on the "most advanced technologies for polysilicon panel production", the energy 'payback' is approximately 6.5 years in Germany[44]. Assuming a life-span for a photovoltaic array of 30 years, this means that about 20% of the electricity produced was 'embodied' up front in the production of the technology - this is an EROEI of 5:1. If the array were to be in production for only 20 years, the EROEI drops to 3:1. In a world of declining fossil fuel production, the first question to be asked is: where is this initial energy to come from? With an EROEI number of less than 5:1, the already optimistic contribution of photovoltaic to global electricity consumption should be derated by 20%, from 0.1% to 0.08%. It is important that the EROEI is kept clearly in mind when talking about energy - as the EROEI drops, less and less of the produced energy is available to use because more and more of it is going back into making the energy-generating technology. In the case of PV, with its already very low contribution to electricity production, derating the contribution for EROEI is lost in the significance of the decimal places.

Sherwani, Usmani & Varun, in a 2009 assessment, suggest payback periods for installed PV that range from 1.7 years to 5.7 years for polycrystalline systems[45]. The related EROEI numbers assuming a 30-year life-cycle are as high as 20:1 and as low as 5:1. The high end of EROEI values would challenge oil and gas as an energy investment in many regions of the world. These numbers vary as widely as they do because of the manufacturing process and the performance of photovoltaic technologies, which depend on the efficiency of the technology and the amount of sun (location) where the system was installed. For example, some technologies take less energy to make (like thin film), but these technologies convert the sun's radiation to electricity at lower levels of performance. If you install the array in a location that does not have day-long sun exposure (if it is shaded by trees or other buildings), if the array is not tilted to the optimum angle at the latitude of installation, or if the array is not pointing due south (in the northern hemisphere), then the PV installation will produce less electricity. If the array does not last 30 years in its productive life cycle, the performance must be derated. If the electricity produced is not used (not tied into a grid or stored using batteries), then then real performance has to be derated, and if you include batteries (or a grid connection, for that matter) these ancillaries should be added to the energy invested in the installation. One also has to be concerned about the location of the manufacturing facility and how far the product must be transported, and what means of transportation (ship, rail, truck) is employed to get the product to the installation. For these reasons, EROEI values are site-specific. It is also for this reason that published EROEI values are typically optimistic.

Since one can't really 'pay back' the energy consumed in producing a photovoltaic array (since the fossil fuel energy has already been burned resulting in emissions of pollutants and greenhouse gases), it is more accurate to think of this (and all alternative technologies) as being a *pollution-diluter.* You are really taking a fossil fuel energy system and making it more efficient (more energy produced overall for the primary energy source burned in the process). To be clear, you could burn coal for electricity at a conversion efficiency of 35% (only 35% of the primary energy embodied in the coal actually becomes energy in the form of electricity). If you use the coal to make electricity, and use this electricity to make PV arrays, then over 30 years you will have produced a net of four times more electricity (at an EROEI of 5:1). This means that you got 40% more energy than what was originally embodied in the coal to begin with. So, you changed a 35% efficient coal-fired electricity generation plant into a 140% efficient coal-fired plant, assuming that all the electricity produced by the plant went into making PV. We won't mention, for now, that it took 30 years to make these gains in efficiency. As a pollution-diluter, we are taking the emissions of coal-fired electricity generation which produces about 975 g $CO_{2(eq)}$/kWh (or natural gas electricity generation which produces about 600 g $CO_{2(eq)}$/kWh) and, over time, diluting the greenhouse gas emissions for photovoltaic systems to a value that ranges from 50 to 250 g $CO_{2(eq)}$/kWh[46].

Even though photovoltaic systems are clearly not pollution-free, they can be a very good diluter of emissions over the productive life when properly installed and operated. The greenhouse gas reduction compared to fossil fuel electricity generation could range from 75% to 95%. It would be truly great if we could manufacture enough photovoltaic systems at a rate that could contribute more than the very small percentage of our massive consumption (0.1%, as discussed above). The challenge to increasing this rate of PV production is that it would require massive investment in manufacturing, which would require stable energy prices or large public incentives. In other words, PV production rates are currently dependent upon the Organizational challenges in an ostensibly 'free' market.

One last concern about photovoltaic: Making everything ultimately has an environmental impact, including pollution in the form of emissions to air and water, and as waste to the landfill. These emissions contribute to ozone depletion, human and eco-toxicity, acidification, not to mention the depletion of raw resources. As a recent Worldwatch report[47] suggests, the increase of photovoltaic production in China has resulted in serious environmental impacts due to the lack of functioning pollution abatement technologies. This example represents the race-to-the-bottom of environmental standards in a globalized economy. Electronics manufacturing is a notoriously polluting endeavor, and it remains to be seen if we are simply trading one type of pollution for another – that is, substituting a reduction of greenhouse gas emissions through photovoltaic production for other serious environmental and social impacts. And given our penchant for off-shoring pollution to developing nations, these impacts should be a much more serious concern for them.

Now Wind ...

Generating electricity from wind energy is considerably more sustainable than photovoltaic for a number of reasons. The energy-return-on-energy-invested (EROEI) is better, the materials used are more common and less likely to be effected by supply, and there are more opportunities for economies of scale.

Wind turbines are largely manufactured from concrete (base), steel (tower and nacelle), copper (generator) and polymers (turbine blade). They typically require steady winds exceeding 20 km/h to generate electricity. Land-based wind turbines can be as large as 3 MW in capacity, though larger turbines are not uncommon for off-shore applications. Many regions of the world have abundant wind, though the existing transmission infrastructure is often inadequate to deliver the electricity to the grid. This is due to the fact that, in the past, most electricity systems have been centralized with the wires extending outward to make a grid, the wires getting smaller and smaller as they spread away from the central plant. The existing centralized generation plants are coal-fired, nuclear and hydroelectric and their locations don't always coincide with the open land (or sea) that might have adequate wind velocities. Obviously, wind farms have to be built where there is ample wind and, as such, the energy cost of wind farms should consider the required infrastructure to deliver the energy to the grid. And since wind is intermittent in energy production, it has been argued that the back-up generation capacity should also be included in the calculations of EROEI for wind power.

A good location for a wind turbine has a *capacity factor* of 35 to 40 percent, which means that the net electricity generated by the turbine over the period of a year is about 35 to 40 percent of the maximum capacity of the turbine. If a turbine has a generating capacity of 2 MW, for example, the average output would be between 0.7 and 0.8 MW at a good location. As the good locations are developed, the capacity factor of wind farms will begin to diminish (not unlike hydroelectric plants after the good rivers are exploited). It is important to keep the difference between 'capacity' and 'delivered energy' in mind when discussing wind energy - the terms are often blurred, and occasionally the terms are blurred in an effort to exaggerate or deceive. To be clear, the 'capacity' of a wind turbine is the amount of power a wind turbine can generate under design conditions. We refer to a wind turbine by this capacity to produce power, 3 MW for example. Remember, a MW (megawatt) is a measure of power (the amount of energy in Joules that can be delivered each second). If a 3 MW wind turbine operated at its design capacity for an hour it would produce 3 MWh of electricity (or 10,800,000,000 Joules). At a capacity factor of 35%, the 3 MW wind turbine can be assumed to be generating power at maximum capacity 35% of the time, and not producing at all 65% of the time. Of course, the actual amount that a wind turbine produces varies with the wind speed, but the capacity factor indicates the average performance compared to the maximum.

Where the deception can occur is when someone says that, say, 600 MW of wind turbine capacity has been installed, comprising 5% of the total capacity in the region. Some governments have targeted that 20% of the total capacity shall be from wind turbines. The problem is that only 35 to 40 percent of this capacity is translated into electricity production for wind, whereas 98% of the capacity of a coal-fired plant is translated into actual production. The 5% of the 'capacity', when it comes from wind turbines, is really 'delivering' only about 1.7% of the demand. An aggressive 20% of capacity might mean that 8% of the demand has been satisfied with renewable wind energy.

Unlike PV (for which there is some predictability, since we generally know when the sun will shine, at what intensity, and for how long) wind is much less predictable. The other advantage of PV is that it typically makes energy when the electricity demand is at its highest. A hot, sunny day when people are working and the air conditioners are blasting coincides with the period when the PV modules will be producing at maximum capacity. Wind turbines, on the other hand, can be producing at capacity in the middle of the night, when demand is at its lowest. What happens to this generated electricity?

The electricity grid must operate within some technical limits to make it functional. A portion of the generation-capacity has to be on all of the time; this is called the base-load. Suitable generation technologies for the base-load include coal-fired, nuclear and hydro. They can operate with a high reliability for long periods of time and, in the case of coal-fired and nuclear, the plants are not easy to turn up and down, on or off. The base-load is roughly the smallest demand over daily and seasonal fluctuations - it is the electricity you can expect to be using all of the time.

Above the base-load, a portion of the power generation must track the fluctuating demand. As noted, coal-fired and nuclear plants have a limited ability to be turned up or down, on or off with the variance of demand. Many grid-systems use natural gas fueled turbines and hydroelectricity to meet these fluctuations, as they can respond quickly to the changes in demand. Some of the gas turbines must be idling (spinning) to be ready to meet sudden changes in demand, while others may be turned on as demand rises. Hydro is a particularly flexible generation technology and responds quickly to demand. PV is a more-or-less predictable technology that produces electricity when it is most required, but wind produces its electricity only when the wind is blowing, which can be at any time, even during a period of low demand. The base-load generation plants cannot be turned off to use the wind power when it is generated, so if the wind turbines are producing when the demand is at levels satisfied by the base-load, this generated electricity has to be wasted (which would lower the capacity factor). The larger the grid, the lower will be the chance of wasting this generated electricity. The point is, however, that if a renewable energy technology produces electricity when it is not needed, it is wasted. The capacity factor should, therefore, reflect utilized capacity (not the percent of time the technology is generating).

As a rule-of-thumb, only about 20% of the system output of electricity can come from intermittent sources like wind, otherwise the grid becomes unstable due to the fluctuation of wind speeds and output. Again, this is 'delivered energy' and this should not be confused with 'capacity' which could easily represent a large portion of the total generation capacity of the system. It is useable output (delivered energy) that matters up to the point where the unpredictability and intermittent nature of the renewable energy destabilizes the grid. The 20% rule-of-thumb is not firm limit, as it depends on the mix of renewable energy technologies, and the size of the region being serviced - but, today, it is a practical value for discussion purposes.

Another question that must be asked is: how much (and what mix of) renewable energy can be used for an electricity grid? We have determined that PV and wind can contribute to the grid (though intermittently). Can a mix of renewable energy sources provide the reliability of base-load? Imagine a calm, cloudy day when wind and PV energy technologies stand idle. In this case, there has to be some alternative energy sources available to meet the demand - unless our demand changes with the availability of energy. There is a story of a couple living on a gulf island near Vancouver who had a small electrical generator in a creek. When it rained, the couple had power and they went to work doing the chores that required electricity. When it was sunny and the creek stopped flowing, they enjoyed the day outdoors. I don't think I'm going out on a limb to suggest that our current economic system wouldn't work well this way (thought it might be nice if it did).

Base-load electricity production commonly comes from the fuels for which we have already developed a massive infrastructure - fossil fuels like coal, oil and natural gas. In addition to fossil fuels, however, base-load may come from hydro where hydroelectric capacity is available (and might be expanded with small hydro and 'pumped storage' opportunities). If stored, it may come from the excess electricity produced previously by renewable technologies (like wind power produced at night). And it may come from bio-fuels like methane produced from the digestion or pyrolysis of organic solids, ethanol from fermentation processes, or waste-to-energy production from burning our garbage.

Storing electricity is still very inefficient and expensive, though there is a lot of experimentation being conducted, including using everyone's electric car as a big storage battery. Bio-fuel production depends on organics that are limited in supply (as will be discussed). Waste-to-energy is just too contradictory - using energy to produce crap that we throw out, and then using the crap we throw out to make electricity that can be used to make more crap. Furthermore, as will be discussed, studies suggest that there is more renewable energy to be realized in recycling compared to combustion, and waste-to-energy puts a high demand on energy waste that must be sustained.

Even though there are some limitations for wind generated electricity, this technology provides some of the cleanest electricity available. Varun, Bhat and Prakash[48] compare fossil fuel electricity generation plants to renewable energy technologies. Their results, like all energy systems, depend on a number of factors like location, the efficiency of the technologies, infrastructure and so on. Wind turbines produce between 10 and 125 grams CO_2 per kWh of electricity generated [g-CO_2 / kWh]. This can be compared to 53.5 to 250 g-CO_2 / kWh for solar PV; 600 g-CO_2 / kWh for gas turbines; and 975 g-CO_2 / kWh for coal-fired generation. Hydroelectricity ranges from 3.7 to 240 g-CO_2 / kWh, and biomass ranges from 35 to 178 g-CO_2 / kWh. Nuclear power is calculated to be 24.2 g-CO_2 / kWh with assumptions around the energy required for decommissioning and managing nuclear waste indefinitely. The emission value for nuclear power has been widely challenged in the literature.[49]

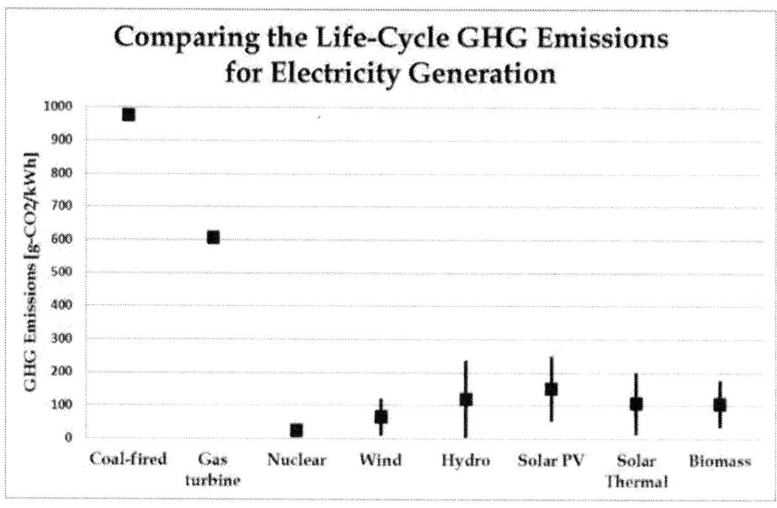

Figure 14 - Life-cycle Greenhouse Gas Emissions

These life-cycle values are based on wind turbines installed in optimum locations over a 20 or 30 year expected operating life span. As noted, once the best sites have been exploited, and due to limitations of urban sprawl and protected lands, the emissions per kWh will generally rise. Furthermore, many turbines will not be operating in 20 years because the advancing technologies will replace the older turbines sitting on the optimum locations. This is already happening and will result in larger emission values over a shorter life-cycle.

As noted with photovoltaic, there are a number of impacts of wind turbines that are not expressed as GHG emissions per unit of energy generated. Wind turbines require non-renewable minerals like alumina (for concrete), iron for steel, copper for the generator, and fossil fuels for the polymer blades. The land required for wind generation is not insignificant (particularly for access roads), and wind farms are known to disrupt the migration of birds and interfere with the biological cues for other species. Furthermore, low frequency noise has been identified as a pollutant for people and other species living nearby. Though it could be easily shown that the environmental impacts of coal-fired plants and hydroelectricity are far worse, this does not indemnify the wind technologies.

Wind is, perhaps, one of the best sources of electricity generation. It is evident, however, that wind technologies can only support a small percentage of the overall electricity demand. The current worldwide production capacity is over 40,000 MW per year, with an installed capacity currently at 160,000 MW[50]. At a generous average capacity factor of 30%, these turbines would generate approximately 420,500,000 MWh of electricity. Remember, our collective consumption of electricity is 20,000,000,000 MWh per year and expected to grow to 25,000 TWh by 2020. As such, wind resources can be converted to produce about 2% of the current global electricity consumption. At the current rate of manufacturing, wind would contribute only 10% of the future *increase* in electricity demand. In other words, we would have to average a manufacturing capacity exceeding ten times the current ability just to keep up with the growing global demand. Due to this large, and likely unrealizable, required investment in manufacturing capacity, it is unlikely wind power will be able to keep up with that demand, let alone replace electricity currently produced from fossil fuels, like coal.

This does not mean the wind is not significant, nor does it mean that some regions couldn't do much better than others in replacing fossil fuel as a source of electricity generation. What it does mean is that it is highly unlikely that wind (and solar) will contribute more than small amount to our current, and increasing, appetite for electric power. And if you consider the dream of converting our transportation system to electricity (electric cars, rail, etc.), this fact becomes even gloomier. The bottom line is that wind turbine and PV technologies will contribute only a small and, likely, declining share to electricity consumption worldwide. This is, of course, if we depend upon market forces. A 'war-time mobilization' of production capacity could change this scenario but, as noted, it is unlikely this will occur until after the energy crisis makes the need to do so patently obvious to everyone, including our leaders. By this time it will be impossible to react meaningfully because we won't have additional fossil energy to invest up-front in the production of renewable energy technologies. Furthermore, all of this is unlikely to occur in a democracy with a functioning 'free' market economy reacting to some combination of energy, economic and social crises.

"From Earth's perspective, however, the American or even the European way of life is simply not viable. A recent analysis found that in order to produce enough energy over the next 25 years to replace most of what is supplied by fossil fuels, the world would need to build 200 square meters of solar photovoltaic panels every second plus 100 square meters of solar thermal every second plus 24 3-megawatt wind turbines every hour nonstop for the next 25 years. All of this would take tremendous energy and materials - ironically frontloading carbon emissions just when they most need to be reduced - and expand humanity's total ecological impact significantly in the short term."[51]

Erik Assadourian

Biofuels

Photovoltaic and wind technologies are used to generate electricity, which leaves an important gap in replacing oil as a liquid fuel. The two main categories of liquid biofuels that may be used to replace petroleum are ethanol and biodiesel, which are derived from many different sources of feedstock. We will first discuss ethanol, then biodiesel.

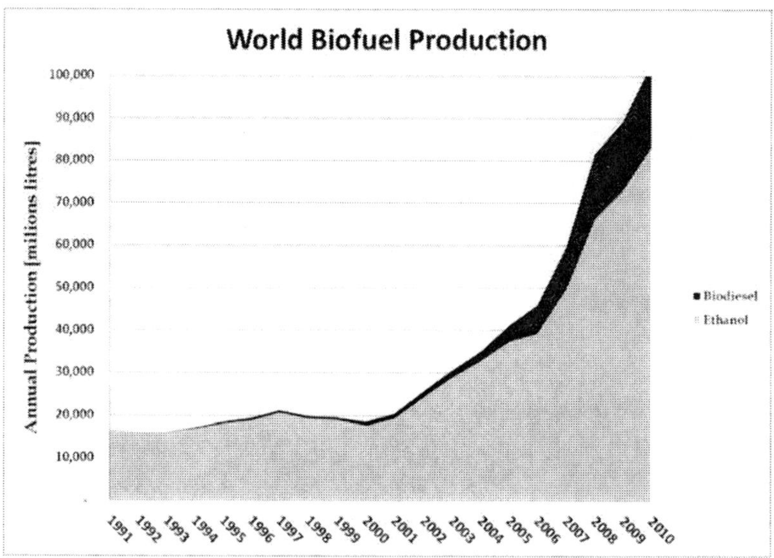

Figure 15 - World Biofuel Production (Biodiesel and Ethanol)

Worldwide ethanol production has grown exponentially since 1975, with a current rate exceeding 80,000 million liters, as shown. Of this global ethanol production, it is estimated that 85% is used for fuel. Brazil and the United States each produce 40% of the global total, Brazil mainly from sugar cane and the United States from corn. China currently produces 5% of the world total, 80% from corn; the European Union produces 2% from grain; and India produces 1% from sugar cane[52].

Like ethanol production, the conversion of vegetable oils or animal fat into biodiesel for fuel has dramatically increased since 1990, as shown. The current global production of biodiesel is roughly 20,000 million liters. The world leaders in biodiesel production are France and Germany, each at 30% of the world total, with the United States producing 12%. Canada produces less than 1% of the world total, with an annual rate of roughly 100 million liters[53].

Though Canada, in comparison to world production, produces insignificant amounts of ethanol and biodiesel, the national potential is significant. If any country could meet their domestic energy needs for transportation using biofuels, Canada is that country: given its relatively small population and large agricultural production capacity.

According to government statistics, in 2008 Canada consumed 1310.8 PJ of energy in 'motor gasoline' for passenger vehicles and for freight each year. At a heating value of 34.8 MJ/liter, this translates to 103.2 million liters of gasoline each day. From the same source, Canadians consume 843.1 PJ of energy in 'diesel fuel oil' for freight and some passenger vehicles each year. At a heating value of 38.6 MJ/liter, this translates to 60 million liters of diesel each day. Using a 5% ethanol blend with gasoline for transportation, the amount of ethanol required would be more than 5 million liters per day. The main first-generation feedstock for ethanol production in

Canada is corn, and 7% of the 9 million tonnes of corn harvested in the country is currently converted into ethanol. The potential for using wheat is also vast considering that 17 of the 20 million tonnes harvested each year are exported[54]. According to Boyle[55], the ethanol yield for corn ranges from 250 to as much as 2000 liters per hectare annually, though other sources suggest yields as high as 3400 liters per hectare. Using an optimistic value for Canadian agricultural of 2500 liters per hectare, the land requirement to meet five percent of the Canadian fuel consumption for transportation would be 750,000 hectares. As such, supplying 5% of our transportation energy requirement with ethanol would require 1.7% of the arable land in the country (assuming a total of 45.1 million hectares), or 65% of the land currently under cultivation for corn.

If wheat were used at a yield of 1000 liters of ethanol per hectare[56] and roughly 8.6 million hectares of land under cultivation for wheat, a 5% ethanol blend would use 22% of the wheat harvested each year. This is based on a replacement of gasoline by volume, so the values would be a little less attractive if one were to correct it for heating value of ethanol which is 29.8 MJ/kg or about 24 MJ/liter (70% of the energy per unit of volume compared to gasoline). Therefore, correcting for heating value, converting all of the corn and wheat in production today into ethanol for fuel would satisfy 20% of the energy demand for gasoline. Converting all of the arable land in Canada to wheat production - all 45.1 million hectares - would only satisfy 80% of the energy demand for gasoline.

Greer provides a similar synopsis for the United States, which shows that less than 60% of the heating value of gasoline could be replaced by ethanol when all of the arable land is put into corn production: "Let's imagine, for example, that the United States decided to replace its current gasoline consumption (a large sector of its fossil fuel use, though not the largest) with ethanol derived from corn. The United States uses about 146 billion gallons of gasoline a year; since ethanol only yields three-quarters as much energy per gallon as gasoline, it would take a bit over 194 billion gallons of ethanol to keep the present American automobile fleet on the road for a year. According to US government figures, there are about 302 million acres of arable land in the United States; corn yields about 146 bushels an acre on average, and you can get 2.5 gallons of pure ethanol out of a bushel of corn. ... the total yield of ethanol would only be a little over 110 billion gallons."[57]

Similarly, Tokar shows that using 100% of the current corn and soybean production will provide only a small percentage of the demand for transportation energy in the United States:

> Even as they [University of Minnesota] project a 25 percent energy gain from bioethanol production, and 93 percent from biodiesel (in contrast to Pimentel and Patzek), they noted that 14 percent of the entire U.S. corn harvest was used in 2005 to produced less than 4 billion gallons of ethanol. This is equivalent to only 1.7 percent of gasoline use, and they also found that 1.5 percent of the soy harvest produced 68 million gallons of biodiesel, or less than 0.1 percent of diesel demand. Thus on a gallon-for-gallon basis, even with a relatively optimistic energy return estimate, they concluded that "dedicating all U.S corn and soybean production to biofuels would meet only 12% of gasoline demand and 6% of diesel demand."[58]

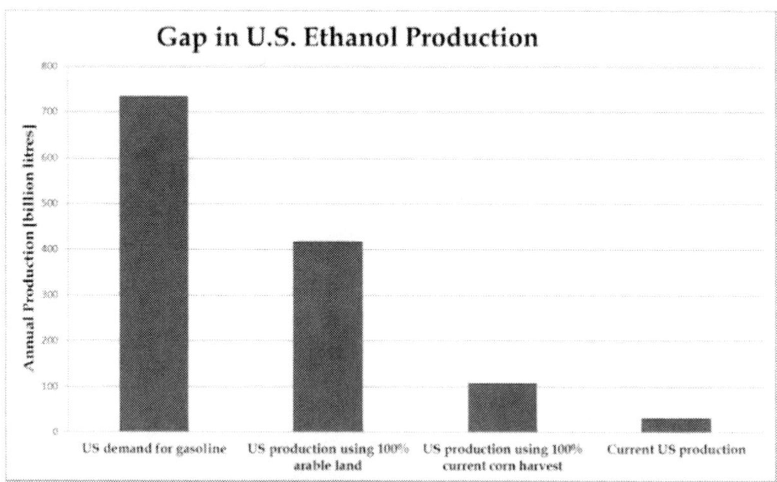

Figure 16 - Gap in U.S. Ethanol Production

It should be noted that these conclusions are based on the assumption that over 3400 liters of ethanol can be produced per hectare of corn production (which is comparable to Pimental and Patzek who used a value of 3200 liters of ethanol per hectare of corn production)[59]. The difference compared to the Canadian yield is due to the estimated yield of 7200 kg/ha in Canada, compared to Pimentel's 8600 kg/ha (at a conversion of 2.7 kg of grain to 1 liter of ethanol).

If that wasn't an Iron-Butterfly-drum-solo-of-calculations … All of these calculations may seem convoluted, but using food grains to make liquid fuel for transportation depends on the yield of the grain, and the conversion from grain to ethanol. In addition to the large land requirements, one should take into consideration the incredible volumes of water used to grow the grains, and one should consider the energy-return-on-energy-invested (EROEI) of the process which makes the use of grain for fuel even less realizable.

It should also be mentioned that there are second-generation feedstock like switchgrass and residuals from forestry that have the advantage of not relying on food crops and can be harvested from degraded or marginal lands. Second generation feedstock using crop, forest and mill residues could be converted into an estimated 21,000 million liters of ethanol each year in Canada (56% of current demand). These projections, however, often ignore the reliance on these lands as a source of forage for livestock, the value of these residues for the health of forests, or land-use by other groups including indigenous populations.

To briefly summarize, if we convert all of the wheat and corn currently grown in Canada, the nation could replace only 20% of its gasoline consumption with ethanol. There are expectations, however, that corn yields in Canada will double over the next decade or two to almost 19,000 kg/ha[60] which would marginally improve the feedstock potential to replace 25% of the country's gasoline consumption. It should also be noted that the current ethanol production capacity in Canada is 1,700 million liters each year, less than 5% of gasoline consumption by volume, and just over 3% by energy. The country is limited by both feedstock and capacity.

Biodiesel production from fuel-crops is likewise limited by available agricultural land for canola production. At 1100 liters/hectare/year, meeting 20% of the 60 million liter/day diesel consumption (a B20 blend) for transportation in Canada would require 40 million hectares, or 88% of Canadian agricultural land. Biodiesel potential in Canada, however, is estimated to be over 500 million liters each year, which is ten times larger than current production rates. Half of this volume may be derived from tallow and yellow grease, over 40% would originate from canola production, and the remainder would come from soy[61]. This rate would satisfy just over 2% of the current demand for diesel.

There is also a large potential for converting biomass to oil, with an estimated 21 million BDt of surplus mill residues and another 29 million ODt of sustainably removable residues (SRR) from crops[62]. At a conversion rate of 150 litres/tonne for biomass using the evolving Fischer-Tropsch process[63], second-generation sources could provide a significant growth potential for biodiesel production: up to 7,500 million liters per year. In other words, if we throw everything we have at biofuel production Canada could meet almost a third of the current diesel consumption for transportation - and like we said, Canada is unusually endowed with huge tracts of land.

On a world-wide scale, assuming that we don't need to grow crops for food, we can convert the 14 million square kilometers of arable land to corn. At a very high average yield of 3000 litres of ethanol per hectare, the world would produce 4,200,000,000,000 litres. Using a value of 24 MJ/l for ethanol, the primary energy available is 100 billion GJ of energy. This is 20% of the current global consumption of 475 EJ. Since we actually do need to use the world's arable land to feed 9 billion people, this is a ridiculous (but telling) thought-experiment.

When evaluating biofuels, it is important to consider the sustainability of the process and the impacts of dedicating arable land for fuel-crop production. As discussed, a useful measure of the sustainability of a process is the energy-return-on-energy-invested (EROEI). This measure identifies the energy required for each stage of the process from sowing seed; the application of chemical herbicides, pesticides and fertilizer; harvesting the crops; transporting the crops to centralized facilities; converting the feedstock into fuel; and the distribution of the fuel product to the consumer. The energy consumed in production may then be compared to the energy value of the fuel when combusted. Because each feedstock is different, requiring different inputs in different regions, and necessitating different conversion processes for varied consumers, EROEI values are difficult to quantify beyond the specific context: the "total global warming footprint depends on what feedstock is used, how and where this feedstock is grown, any land-use changes, and how the fuel is processed"[64]. As such, the following comparison of the EROEI for different feedstock is highly abstracted.

The global average EROEI for the conversion of first-generation feedstock for ethanol ranges from 1.5 for corn, 2 for wheat, to as much as 8 for sugar cane[65]. Second-generation cellulosic feedstock have the benefit of reducing or eliminating the application of fossil-fuel based herbicides and fertilizers, as well as the energy required for seeding in the case of perennial grasses. The EROEI for switchgrass is estimated to be as high as 4.4[66], and experimental processes indicate a potential EROEI as high as 36, rivaling most current energy sources. As a generalization, biofuels would provide greater overall benefits if the feedstock were produced with low agricultural inputs (i.e., less fertilizer, pesticide, and energy), produced on land with low agricultural value, and required a low energy input to convert the feedstock to biofuel[67].

The EROEI for biodiesel production is similar to ethanol production with values of 2.5 for canola, and 3 for soybeans in North America, and values as high as 9 in other regions depending on the feedstock[68]. Biodiesel derived from tallow (animal fats) has an EROEI of approximately 4.5[69], though this value is sensitive to methodology and assumptions, as tallow is considered a low-value byproduct and the calculations do not consider the energy invested in the livestock (raised for meat as a primary product).

There are experiments in using algae grown in conditions that can continually produce biofuels (as opposed to the seasonal constraints of crops). High costs and a currently very low EROEI values remain barriers, and the absence of commercial-scale production makes this, at best, a long-term possibility for a civilization determined to jeopardize its existence in the short term.

Beyond the energy balance for biofuels, social issues like food-for-fuel impacts on the cost of staple grains worldwide, the quality of life for farm labour, and the social health of rural communities must be considered. Impacts like the eutrophication of watersheds and oceans due to uncontrolled runoff of the fertilizers applied to fuel-crops; the contribution to global warming of nitrogen compounds from fertilizers and the destabilization of soil; erosion from unsustainable agricultural practices; biodiversity loss from land conversion; and the consumption of non-replenished water from deep aquifers should all be evaluated. As Tokar has suggested:

> Agrofuel crops also put a significant strain on water resources, from expanding corn crops in the U.S. to cane fields in Brazil. Sugercane ethanol, for example, requires 2,200 gallons of water for every gallon of fuel that is produced. For corn ethanol produced in the U.S., estimates range from 10 to 1,000 gallons of water per gallon of fuel, ... Researchers in the U.S. and overseas have raised alarms about the increasing demand on the world's scarce irrigation water as agrofuel production increases, and fertilizer runoff from increased corn acreage for ethanol is also contributing to the expansion of the Gulf of Mexico's (pre-BP oil spill) hypoxic "dead zone."[70]

It has also been argued that land-use change from peatlands and grasslands to biofuel production may "create a 'biofuel carbon debt' by releasing 17 to 420 times more CO_2 than the annual greenhouse gas (GHG) reductions these biofuels provide by displacing fossil fuels"[71].

> Following this approach, they concluded that it would take 167 years for the savings of CO2 emissions from using ethanol to compensate from the initial land use conversions, and that for the first thirty years, "emissions from corn ethanol [are] nearly double those from gasoline." Even with more optimistic estimates of future corn yields and new conversion technologies, the thirty-year emissions would continue to exceed those for gasoline. The use of switchgrass instead of corn to produce ethanol would require a fifty-two-year payback and results in a 50 percent increase in emissions over thirty years; however sugar cane grown on former grazing land "could pay back the upfront carbon emissions in four years."[72]

The bottom line is that the United States, and most other countries in the industrially developed North, cannot meet more than a tiny fraction of the current energy demand in the transportation sector with biofuels: Mathews agrees with the "view that the US (and OECD countries more generally) cannot meet their transport fuel requirements through production of ethanol and biodiesel from domestic biomass sources" and suggests that these countries should "embark on a massive import program involving ethanol and biodiesel imported from countries of the South where the conditions permit it." [73]

Fuel crops that can be grown in the South, like sugar cane, offer vastly higher yields than corn, and crops like miscanthus offer comparative yields on marginal land with multiple growing cycles each year. Africa and South America have vast areas of arable land not under cultivation and, in the infamous words of Lawrence Sumners: "I've always thought that under-populated countries in Africa are vastly under polluted" and, by extension, under-exploited. Mathews offers a vision of a mutually beneficial relationship, or biopact, in which the South is provided assistance to develop industrially while providing the North with the biofuel resources it requires for transportation. The Clean Development Mechanism (CDM) of the Kyoto Protocol provides a template for this. More likely, however, the result will be a form of eco-imperialism in which countries of the North directly exploit the production potential of tropical countries through traditional levers like the World Bank and IMF, or an extension of the war-for-energy-security stratgy; perhaps eco-imperialism will become another dimension of Bricmont's 'humanitarian imperialism' in which personal interest can be aligned with feel-good moral imperatives[74]. Regardless of the mechanism, the new cash-crop for the global South will be fuel. Distant arable lands will be mined for energy, which will widen the metabolic rift between North and South.

Competition for land that would otherwise grow food or provide for indigenous peoples still living on the commons is a serious concern. A 2008 Oxfam report

> considered the global implications, concluding that even the entire world's current supply of carbohydrates, including all starch and sugar crops, could replace at most 40 percent of petroleum consumption, and all the world's oilseeds would displace less than 10 percent of the world's diesel fuel. In early 2010, the British anti-hunger charity Action Aid

> projected that global biofuel targets could raise food prices an additional 76 percent by 2020, increasing the number of hungry people in the world by an estimated 600 million. Using a relatively conservative figure of 30 percent of recent food price rises attributable to fuel crop production, they estimated that 30 million people are now going hungry due to agrofuels, with a further 260 million people at risk of hunger.[75]

To summarize, there is not enough arable land to produce even a fraction of the energy required for transportation in the world. Biofuel production today accounts for less than 2% of the global demand of oil-energy, and less than a half of one-percent of the total global energy consumption. EROEI values remain low and, as a result, the net reduction in greenhouse gas emissions is small (though there is some argument that second-generation biofuels will be much better). The development of second-generation biofuels is also hampered by costs of processing and fluctuating fossil fuel prices. Worldwatch suggests that capital investment costs for second-generation biofuel plants "are thought to be three to four times those of first-generation biofuel plants, and production costs are higher as well."[76] Volatility in fossil fuel prices also makes these investments insecure: As prices for fossil fuels increase, investment is encouraged, but as fossil fuel prices collapse from the recession caused by previously high fuel prices, the biofuel plants close their doors. It doesn't take more than a few cycles to discourage private investment in the long term.

"In such a future, the periods of apparent recovery that will likely follow each round of energy shortages and demand destruction will provide little room to rebuild what has been lost. Those periods will, however, make it exceptionally difficult for any response to fossil fuel depletion to stay on course, so long as that response relies on market forces or politics for its momentum. Each time oil prices slump, the market forces that support investment in a sustainable future will slump as well, while governments facing calls for limited resources will face real challenges in maintaining a commitment to sustainability, which for the moment, no longer seems necessary."[77]

John Greer

Clean Coal? Go, Go, Gadget, Nuclear?

I don't believe in illusions
'cause too much is real.

Johnny Rotten

I'm not sure that it is worth talking about so-called 'clean coal', or the illusions of nuclear power, for that matter. They are acts of desperation, at best.

Clean coal - an oxymoron, if there ever was one - is premised on capturing the emissions from coal burning and disposing them in some permanent way: carbon sequestration and storage. The technology is available to do this, even if the permanent storage remains a problem. Much of the carbon dioxide from the emissions of a coal-fired electricity generation plant can be captured (for new plants), compressed, and injected deep underground in saline aquifers. It takes energy to do this, of course - some suggest as much as 60% more, making the net efficiency of the generation plant drop from 35% to around 20% (though the net emissions of CO_2 are much less).

The main problem is that there is no clear understanding if the carbon dioxide will stay underground, and how much the underground aquifers can absorb over time. A commercial scale experiment near Weyburn, Saskatchewan seems to have failed with carbon dioxide bubbling up on nearby farms from the supposedly secure storage underground. Clean coal also doesn't address the many other environmental and social impacts of mining coal, and disposing of the toxic ash.

Hoggan gives the following summation of a clean coal plant:

> Then this "clean" coal plant produces the following toxins: 10,000 tons of sulfur dioxide, the main component of acid rain; 10,200 tons of nitrogen oxide, a major cause of smog and contributor to acid rain; 3.7 million tons of carbon dioxide; 500 tons of small particles, a major contributor to lung disease; 220 tons of hydrocarbons, smog-producing particles of unburnt fuel; 720 tons of carbon monoxide, a greenhouse gas that is also poisonous to humans; 125,000 tons of ash and 193,000 tons of sludge form the smokestack scrubber. The ash and sludge consist of coal ash, limestone, and many pollutants, including toxic metals such as lead and mercury (...); and 225 pounds of arsenic, 114 pounds of lead, 4 pounds of cadmium, and many other toxic heavy metals.[78]

As for nuclear, where does one start? Okay, from the beginning. Uranium ore must be mined, and though there is a lot of uranium on earth - even in sea water - the grade of the ore makes most of it energy prohibitive to extract. As a result, uranium has the same problem of coal, oil and gas - finite reserves, peak extraction capacity of uranium and diminishing EROEI over time: "Another way of putting it is to say that if all of the electrical energy used today were to be obtained from nuclear power, all known useful reserves of uranium would be exhausted in three years."[79]

This fact alone is enough to make the idea that nuclear power can fuel the future a non-starter. But if we were to go on, one would have to mention the negative environmental impacts of mining uranium and the health impacts on the workers and people living downstream and downwind of the tailings[80]. One would have to mention the risk of proliferation of nuclear weapons, and the misuse by 'rogue' groups in an increasingly destabilized world, including the greatest nation that ever bombed the earth, the United States, who continue to claim the (unilateral) right to use nuclear force to secure its interests.

Furthermore, economically, nuclear can no longer compete with safer technologies - Thomas Homer-Dixon has recently stated that "the capital costs of nuclear plants have skyrocketed, with estimates of the final price of plants under construction in Europe and North America coming in three to four times above initial projections"[81] As Amory Lovins has said: "Coal-fired and nuclear and gas-fired central power plants are all dying of an incurable attach of market forces ... Basically what happened is that central plants cost too much and have too much financial risk to be attractive for private capital ... That's why, in 2006, [the nuclear industry] worldwide installed less capacity than solar cells added; a tenth as much as wind power added, thirty or forty times less than all forms of micro-power. Just distributed renewables got US$56 billion of private risk capital, nuclear got zero; it's only bought by central planners. There are no orders in the United States, despite subsides now roughly equaling or exceeding the total cost of the (nuclear) plants."[82]

And we would have to build two or three thousand 1 GW nuclear power plants in the world to replace coal[83]. Homer-Dixon summarizes the situation:

> We're in an energy box. The walls are high and thick, and they're closing in. The main source of the energy that drives our civilization is not viable in the long term. Eighty per cent of our energy comes from carbon-based fuels, and their emissions are wrecking our climate. But every direction we turn to get out of this

> box seems blocked by technological, economic or political obstacles.
>
> Storing the emissions of carbon-based fuels underground is phenomenally expensive. We've already dammed most of the best hydropower sites. Renewables such as solar and wind are too intermittent and diffuse to supply more than 20 per cent to 30 per cent of our needs. Biofuels such as corn-based ethanol take nearly as much energy to make as they give back. And nuclear power scares the wits out of people and is, anyway, pricing itself out of the market. [84]

Clean coal and nuclear power are attractive mainly because of their ability to provide base-power to the electricity grid - something that renewables cannot do without massive investment and extremely efficient storage, which doesn't exist. Another problem is that they are not really a short-term solution (due to the time it takes to build these plants), and they are not really a long-term solution due to the limits of sequestering carbon emissions in the case of clean coal, and the finite reserves and diminishing EROEI of the fuel sources in both cases.

Waste

'Man is distinguished from all other animals by the limitless and flexible nature of his needs.'[85]

We have discussed the enormous quantity of resources we consume for our way of life. Freudenburg presents "that, excluding water (which would make these numbers far higher), the total inputs to the U.S. economy in 1993 amounted to roughly 20.8 billion metric tons of raw materials. The outputs of the economy, on the other hand, included just 0.5 billion metric tons of products (508 million metric tons to be precise) - and the number is only that high because it includes 200 million tons each of exports and of food. ... The U.S. economy also produces some 2.8 billion tons of durable goods, such as factory machinery and refrigerators, and more importantly, new construction: ... The other 18 billion tons or so (about 90 percent of all raw material inputs) got turned into wastes and emissions by producers, all before any goods were delivered to consumers."[86] As such, "One half to three quarters of annual resource inputs to industrial economies are returned to the environment as wastes within a year."[87]

Since 90 percent of the waste is created before the consumer receives the product, only a small percentage of the total waste from extracted resources results in consumer waste (as substantial as that might be). "Bottom line: it would be great if consumers recycled everything they purchased - that 100 million tons of actual products - but even if they did, they would be recycling less than 1 percent of the environmental resources actually used in the U.S. economy."[88] In other words, most of the waste is created in the process of mining and manufacturing, and only a small percentage of the resources extracted to make our products are actually embedded in the product, and only a portion of the products consumed are recyclable. The dream of zero-waste in a municipality (dealing with consumer waste) will have no substantial impact on conserving resources.

This is supported by Foster who argues that: "All too often garbage is treated as a problem mainly associated with the direct consumption of consumers. But municipal solid waste in U.S. society is estimated to be only some 2.5 percent of the total waste generated by the society, which also includes: (1) industrial waste, (2) construction and demolition waste, and (3) special waste (waste from mining, fuel production, and metals processing)." And he quotes Derrick Jensen and Aric McBay who observe that: "If we divide municipal waste by population, we get an average of 1,660 pounds per person per year. But if we include industrial waste, per capita waste production jumps to 16.4 tons per person, or 52,700 pounds."[89]

Really?! Do we each really produce almost 5 pounds (over 2 kg) of waste each day by discarding our consumer goods? And are we accountable for over 90 pounds of waste each day if industrial waste is included in the calculation? What compels each of us to waste so much? Foster argues that capitalism (our Organization of the economy) is premised on consumption and waste which is driven by low prices and good marketing that create artificial use values as a substitute for genuine need:

> The result is the production of mountains upon mountains of commodities, cheapening unit costs and leading to *greater squandering of material resources*. Under monopoly capitalism, moreover, such commodities increasingly take the form of artificial use values, promoted by a vast marketing system and designed to instill ever more demand for commodities and the

> exchange values they represent - as a substitute for the fulfillment of genuine human needs. Unnecessary, wasteful goods are produced by useless toil to enhance purely economic values at the expense of the environment. Any slowdown in this process of ecological destruction, under the present system, spells economic disaster.[90]

Herman Daly makes a similar argument that our 'relative needs' - those needs that are manufactured by cultural expectations and manipulations - are insatiable:

> Keynes (1930) argued that absolute wants (those we feel independently of the condition of others) are not insatiable. Relative wants (those we feel only because their satisfaction makes us feel superior to others) are indeed insatiable, for, as Keynes put it, "The higher the general level, the higher still are they." Or, as J.S. Mill expressed it, "Men do not desire to be rich, but to be richer than other men." ... Since the struggle for relative shares is a zero-sum game, it is clear that aggregate growth cannot increase aggregate welfare."[91]

Consider that since 1970 "the size of a new house has gone up about 50 percent, electricity consumption per person rose more than 70 percent, and municipal solid wastes generated per person went up 33 percent. Eighty percent of new homes since 1994 have been exurban, and more than half the lots have been ten acres or more. Yet even the larger homes and lots are too small to contain all the accumulating possessions. The self-storage industry didn't begin until the early 1970s but has grown so rapidly that its buildings now cover more than seventy square miles ..."[92]

The argument that we waste because of a disconnect between genuine need and manufactured need is supported by the statistics on who does most of the wasting: "An analysis prepared for the *1998 Human Development Report* found that 20 percent of the world's people in the highest-income countries account for 86 percent of total private consumption expenditures, 45 percent of meat and fish consumption, 58 percent of energy consumption, 84 percent of paper consumption, and 87 percent of the world's vehicle fleet."[93] And remember that our satisfaction of meeting basic needs is achieved at about $15,000 of purchasing power - for those with more disposable income, consumption and the concomitant waste does not serve greater satisfaction. What does it serve? It serves the need of capital to perpetuate growth, but it does not serve the need for sustainable human civilization or the earth itself: "Too easily forgotten is Gaia's need: we have to leave enough natural ecosystems on land and in the ocean for planetary self-regulation."[94]

Using Waste to Make More Waste

It is called 'Waste-to-Energy'. The context is that we have a lot of waste and we currently bury most of it in the ground (or dump it into our oceans, or simply incinerate it). But what if we burn this waste to make energy?! It's the perfect solution - using waste to make energy so we can make more products that we don't need, so we can throw them away to make waste that can make more energy to make new things we don't need, dot, dot, dot.

As previously noted, North Americans produce 2 to 3 kg of solid waste each day (roughly, our bodyweight each month). This pales in comparison with the weight of carbon dioxide we send into the atmosphere each day, but in this case we can see it and we have to move it out of our sight. Of this solid waste, much of it is recyclable or compostable - it can be put back into the materials cycle or back into the earth to replace the nutrients that we mine from the soil for our food. That would be the sensible solution to managing our waste, but we only do this for about a quarter of our consumer waste - why is that? It seems that we are willing to spend money to build infrastructure to burn it to make energy, but we're not interested in doing the much more logical act of recycling or composting.

Whether recycling and composting or converting it into energy, it is better than putting our waste into the landfill - that's truly a waste. It is a waste of raw resources and it is a waste of energy. Worse, because we are in the habit of putting everything into our waste stream - chemicals, batteries, and other potentially hazardous wastes - when it sits in the landfill for years it creates a toxic brew called leachate. The leachate is the water contained in the waste stream or rain percolating through the waste that settles at the bottom of the landfill. In newer landfills this leachate is contained by an impermeable liner, from where it may be collected and treated - though most treatment systems do not remove many of the synthetic chemicals and heavy metals present in the leachate before it is diverted back into the ecosystem. But for most landfills, there is no liner or the liner has been damaged and the leachate just keeps on percolating into our groundwater. Not only have we wasted our raw resources by burying them, we create a toxic mess that contaminates the fresh water that we rely on. Landfills also act as big composters where organic materials decompose in the presence of water and oxygen. Some landfills are dry, so this process does not occur, but when water and air are present, lots of carbon dioxide and methane gases are produced. Landfills are a major source of greenhouse gas emissions in the world. Landfills ... not a very smart idea. But they are cheap, and we like cheap - even if we are simply avoiding the real costs until a future time (expensive water treatment, health effects, climate change, and so on). So, we agree that landfills are not our best choice for dealing with our wastes. And we don't seem to care much for recycling. What are the alternatives? How about burning it for energy?

Currently employed or emerging technologies for waste-to-energy include incineration, fluid bed gasification, pyrolysis, and plasma arc gasification[95]. Incineration is old-school. It is simply burning the waste in single stage or multiple stages. Traditional incinerators do not separate the waste before incineration which results in dangerous air emission. The thermal energy captured in the system is usually used to satisfy nearby heating or process demand. Electricity generation is not typical for incineration facilities, though solid waste may be combined with coal or other solid fuels in other power-generation processes.

For emerging technologies the waste is typically pre-treated to separate out compostable and recyclable materials and then shredded and packaged in bundles with a consistent mix of waste that provides the required heating value for the process. The waste is then either burned as a fuel, or burned with restricted volumes of air to produce combustible gases or oil which can then be combusted as a source of energy. In a combined cycle system, the combustible gases or oil may be burned to make electricity and the waste-heat generated can be used to heat buildings or nearby processes. In well operated systems, pollution control devices remove the heavy metals, like mercury, and synthetic chemicals in the air emissions. One of the worst pollutants created in incineration process is dioxin, which is perhaps one of the most toxic chemicals we humans have managed to create. The ash produced in the process contains the residuals of the waste that hasn't burned, and it often has concentrated amounts of hazardous heavy metals and chemicals. Many processes will make the effort to stabilize these hazardous elements before diverting them to the landfill (oh, yes, we still need landfills).

On the face of things, waste-to-energy seems like a win-win approach to solid waste: one can reduce the amount of waste being dumped into big, expensive holes in the ground while making energy (and money) at the same time. On the positive side, waste-to-energy technologies may reduce the amount of greenhouse gases compared to landfills, by burning the waste to create carbon dioxide compared to the more potent greenhouse gases emitted by landfills. It has been estimated that waste-to-energy reduces greenhouse gas emissions by 1 tonne per tonne of waste not sent to a landfill.[96] They can also reduce the amount of waste going to the landfill by 90% (by volume) or 75% (by weight). And the technologies can make as much as 500 kWh of electricity for every tonne (1000 kilograms) of waste incinerated.

How much energy is that? Let's take the value of 418 kg/person each year of waste. There are a few ways to look at this, but let's say there are 2.5[97] people on average in every home – that makes 1.05 tonnes per year per home. At 500 kWh, this waste can produce 550 kWh per home each year – or almost 10% of the average electricity used by the same home in North America. Of course, this does not include the electricity used by industry and business. To see this in another way, let's look at a particularly affluent and, therefore, wasteful Canadian province – Alberta. Albertans (approximately 3.7 million people) produced almost 1000 kg/person of solid waste and consumed over 70,000 GWh of electricity in 2010. That means that Alberta could produce 2.6% of their total electricity consumption using waste. On a larger scale, Canadians produce 13.4 million tonnes of garbage each year with about 27% being recycled. Using a heating value of 6 GJ/kg (for MSW after recycled materials are removed), waste could provide about 1.3% of the country's electricity. As a primary energy, municipal solid waste would deliver less than 1% of the total energy demand in a country that is particularly good at making waste. If only we could waste much, much more ...

On the positive side, waste can be used to make some energy. On the negative side, these emerging waste-to-energy processes are capital intensive and expensive to build and operate (the energy produced does not offset the costs at current energy prices). The ash produced can be very toxic depending on the care taken to separate hazardous materials from the waste stream before burning it, and the methods used to stabilize it before disposal. And, of most concern, there is a need to feed the machine once it has been built. In other words, the waste stream must be maintained. This means burning materials that should be recycled, either because it more convenient to do so (compared to the sorting, handling and transportation of recyclable materials), or because the waste stream diminishes for some reason and the flow to the waste-to-energy process has to be maintained. A well designed and operated waste-to-energy system can support recycling, in principle, by sorting it from the waste stream before the remainder is burned. But in the absence of reprocessing facilities for recycled materials, the tendency will be to burn it all (and there is a lot of energy in plastics, paper, and cardboard).

The lowest possible environmental impact of waste management systems is premised on high levels of separation without incineration. The impact evaluation of total environmental impact is based on 'renewable energy use, total energy use, water, suspended solids and oxydable matters index, eutrophication, and hazardous waste"[98]. Life-cycle assessments[99] for solid waste suggest a hierarchy of options from avoidance/reduction, to reuse, to recycling, to waste-to-energy, and lastly, to landfill. The major problem with waste-to-energy technologies is that it contradicts the most effective way to deal with waste – that is conservation. In a way, waste-to-energy is a cynical (like I should point fingers) act that has already decided that people will not conserve and that they will continue to produce lots of useful waste into the foreseeable future. In other words, waste-to-energy is banking on the observation that there is absolutely no way that we will respond meaningfully to the many challenges facing civilization by reducing our consumption. I agree, but thought that I would point out the contradiction anyway.

Green Building Instead of Good Building

We've spent a lot of time explaining why we are unable to replace fossil energy with alternative sources. The main issue is that we use an enormous amount of energy, and the alternative sources have much a lower EROEI potential, not to mention a very limited manufacturing capacity for the alternative technologies at present. Perhaps the solution will be conservation (?) The two most significant sectors of energy consumption are transportation and buildings. Let's look at buildings to see if we can put a dent in our fossil energy dependence.

The largest consumers of energy in North America are our buildings: heating, cooling, and lighting. As much as 40% of the total energy consumed is directed to buildings, with another 33% for transportation. As such, one would think that reducing the impact of our buildings on the environment would be a major focus of municipalities and builders. Compared to transportation, buildings will be around for a long time. You could practically replace the whole transportation sector with high efficiency technologies in a decade, whereas our buildings will be around for 60, 80, 100 years or longer. This means that the crap we are building today will have negative impacts long into the future. Even though we know that there are energy and environmental challenges bearing down on us all, our legislation, codes, specification and building practices remain largely inert.

Well-built homes, offices, institutions, commercial spaces and industrial buildings can be designed to reduce energy consumption by 60 to 80% with known and available materials and technologies. Why are we not taking advantage of them? There are a number of reasons[100]: discounting future gains compared to up-front costs; uncertainty around energy prices;

a fascination with sexy technologies at the expense of using more durable and effective materials used in the, yawn, building envelope; confusion about best practices which are not generalizable across many climatic regions of a country; point chasing in certification systems for green building; perverse incentives; and, last but not least, a wide-ranging ignorance in industry that spans builders, realtors, facility operators, to the buyer.

Perhaps the most insidious reason for not building green is the discounting of future gains. People buying a home, for example, might be offered an upgrade in insulation or windows to reduce heat loss. The additional cost is required up-front in the building price, and can be added to the mortgage. The astute buyer-as-investor will then look at the fiscal return for the additional costs compared to long-term bonds, or fanciful gains on the stock market. Since money is being saved in the future operation of the building, it can be discounted at this alternative rate, typically making it unattractive as an investment. This can be exacerbated, when the future gains are based on currently low energy prices - future prices being pure speculation. This logic is applied a bit narrowly, of course, since the same buyer-as-investor will also 'invest' in the Jacuzzi tub, the old-growth hardwood flooring, entertainment rooms, and any number of other luxuries. These will add 'value' to the price of a home, it is believed, whereas good materials and performance are not perceived in the same way.

Basing decisions on future savings from better performance does not readily consider costs externalized to society, and it does not consider improved home comfort and better air quality for better health which have societal cost implications. It also does not consider that if everyone pursued the better 'investment' we would not be able to justify much of the infrastructure used by our society today, which is often financed with low-yield bonds. In reality, we should be happy as home owners to get our money back over the life-cycle of the home, which is not really an investment, but a place where we live and raise our families. So, point one: Jacuzzi tub trumps insulation as an investment.

Another problem with green building is that technologies are much more appealing to the buyer than good building materials and practices. Most of this can be attributed to social prestige - the photovoltaic array on the roof is much more impressive than the high performance windows that look just like every other type of window. Be honest, for most of the people you know, which would they chose - a Mercedes body with a Chrysler Neon drive train, or a Toyota Corolla body with a Porsche hybrid drive train? How many people feel a stirring in their loins when they see durable roofing, or insulated concrete forms used for the basement wall? So, people will build ridiculously large homes on what was once good farmland, leave on the lights and computers, run a 152 inch plasma television, and then buy a 10 kW photovoltaic system to run it all when half of the power being generated is used for wasteful behaviors. In the final analysis, one is simply *consuming to reduce a consumption problem.*

The third deterrent for green building relates to the generalizability of best practices. To be green, a building has to be designed for the local climate. The optimum insulation, shading, colors, heating and cooling systems, water conservation devices, and so on depend largely on the location. A home designed for optimum performance in Chicago will differ from a home built for Los Angeles or an arctic community like Akviligjuaq, Nanavut. A German Passivhaus or a Brittish cob house may not perform so well in Saskatoon, Canada. So, going to the bookstore and buying a book on green designs may be confusing and even misleading, as the authors tend to generalize the benefits which may or may not be appropriate for every location.

This criticism may be extended to green rating systems designed to certify achievements in green building. There are many rating systems and certification process being used around the world, but most are prescriptive - do this and get these many points - regardless of the local efficacy. To take one example, a local builder in my community achieved the highest rating in a green building rating system for using a number of technologies including a ground-source heat pump ('geothermal') for heating and cooling. This system uses electricity to operate a refrigeration cycle that employs the relatively constant temperature of the earth to heat and cool the home. In principle, the system uses one unit of energy in the form of electricity for about 3.5 units of energy in the form of heat. That is an EROEI of 3.5 on the surface of things. But if one goes back to the primary sources of energy, one soon finds that the electricity was generated by a coal-fired plant with a 30% thermal efficiency. With this in mind, the EROEI based on primary energy is fast approaching break-even. Now if one considers that coal-fired electricity is much more polluting than natural gas, and that natural gas can be burned for the same application of home heating at an efficiency of over 90%, it can be concluded that the ground-source heat pump was overall more costly to buy and operate, requires more maintenance, and creates more pollution than a high-efficiency gas furnace. But the home received mucho 'green' points for installing the system and, even more perversely, the government granted an incentive to install the system (and ground-source heap pumps represented the only green incentive the government was offering at the time for green building). This story would be completely different in a region where electricity was greener.

To summarize, point-chasing in green building rating systems is not always green, as it depends on the characteristics of the location (climate, energy sources, water sources, potentials for alternative energy, building codes and architectural controls, etc.). *What you measure is what you get,* so if you are giving points for something you will get that something - even if it wasn't what you expected.

And, finally, green building doesn't work because realtors, builders, and the buyers are quite often illiterate as it relates to energy, water, and air quality. How does a builder decide to build green when they cannot optimize (which requires some energy modelling)? How does a realtor sell the value embedded in a green home when they have no idea how to translate the better performance into value? How does a home buyer differentiate between the yummy Jacuzzi tub from a triple-pane, low-e, argon filled, fiberglass-framed window with a low-solar-heat-gain coating? Adding to the confusion are the shysters trying to sell magical technologies to the public; the lobbying to government for incentives for their industries (with the promise of green jobs); a for-profit-driven building industry competing for contracts; and the ongoing dissimulation from energy corporations.

... we couldn't do anything about it if we did (continued) ...

Like photovoltaic and wind, biofuels can be reasonably expected to provide only a small fraction of the energy we require to maintain our current population and lifestyle. These renewable energy technologies are not without environmental and social impact, and they do not eliminate greenhouse gas emissions. The infrastructure required to increase the conversion of renewable energy sources is massive and the required increase in capacity will demand an up-front investment of fossil fuel energy. As the rate of fossil fuel energy production is in decline and as it takes more energy to make energy (declining EROEI), the energy required to be invested in processing and infrastructure will necessitate that another demand for the energy remains unmet. It would take a leap of faith to believe that the blind 'invisible hand' will know where to invest this energy.

A recent IPCC report on renewable energy focuses on the global 'potential' for using renewable energy sources to replace fossil fuels.

> As infrastructure and energy systems develop, in spite of the complexities, there are few, if any, fundamental technological limits to integrating a portfolio of RE technologies to meet a majority share of total energy demand in locations where suitable RE resources exist or can be supplied. *However, the actual rate of integration and the resulting shares of RE will be influenced by factors such as costs, policies, environmental issues and social aspect.*[101]

There is no doubt that there is a lot of sunlight, wind, and plants to make energy with. The key, as presented here, is that though the potential is high, the rate of integration of renewable technologies can be severely limited by energy-investment in infrastructure and processing capacity (EROEI), social and environmental impacts, and economic ideology. I would like to repeat Assadourian's comment, as he has correctly encapsulated the issue best:

> From Earth's perspective, however, the American or even the European way of life is simply not viable. A recent analysis found that in order to produce enough energy over the next 25 years to replace most of what is supplied by fossil fuels, the world would need to build 200 square meters of solar photovoltaic panels every second plus 100 square meters of solar thermal every second plus 24 3-megawatt wind turbines every hour nonstop for the next 25 years. All of this would take tremendous energy and materials - ironically frontloading carbon emissions just when they most need to be reduced - and expand humanity's total ecological impact significantly in the short term.[102]

The fact of the matter is that we don't have the production capacity we require, and an increase in this capacity would require additional fossil energy to 'invest', political vision, massive financial investment, and an economy conducive to a change in infrastructure - none of which we have.

The Jevons Paradox (The Rebound Effect)

One of the more interesting relationships as it pertains to our vain hopes to dematerialize our society, or reduce resource consumption through gains in efficiency has been called The Jevons Paradox (or the Rebound Effect). Jevons was a British economist who, at the end of the nineteenth century, "argued that increased efficiency in using a natural resource, such as coal, only resulted in increased demand for that resource, not a reduction in demand. This was because such improvement in efficiency led to a rising scale of production."[103] Moreover, Jevons wrote: "*It is wholly a confusion of ideas, ... to suppose that the economical use of fuel is equivalent to a diminished consumption. The very contrary is the truth.* As a rule, new modes of economy will lead to an increase of consumption according to a principle recognized in many parallel instances."[104]

The Jevons Paradox is reflected in many of our behaviours. For example, "the fuel efficiency of cars in the UK has improved over the last decade, saving the equivalent of 0.5 mtoe (million tonnes of oil equivalent) per year. But at the same time the ownership and use of cars and the distance they travel has increased the equivalent energy use by 0.9 mtoe per year."[105] You would not be hard pressed to find similar economic responses to any energy saving or efficiency-gaining technology. Take LED lights as an example. LED lights offer the consumer a 99% reduction of energy use with a fifty times longer lifespan. The cost difference between the two has dramatically decreased to about 10 times more for the LED than the traditional incandescent light. This should mean that our electricity consumption will drop noticeably with the greater performance. But (in line with Jevons' observations) this will not be the case, because this long-life, low-energy light can now be used to illuminate areas that would never have been considered before. Now we are lighting the whole sides of buildings with advertisements - even the underside of park benches for aesthetic purposes. Many people have installed them in the refrigerators, because the one that comes with the appliance keeps going out when the door is closed (just kidding ... I think).

This paradox has profound implications for efforts to conserve natural resources like fossil fuels. "For example, the development of renewable energy resources, such as wind and solar power, are commonly identified as a way to reduce dependence on fossil fuel, based on the assumption that the development of alternative sources of energy will displace, at least to some extent fossil fuel consumption."[106] If you consider alternative energy technologies as simply making fossil fuels more efficient (as argued above), this efficiency will only exacerbate fossil energy use as the prices for fossil energy drop with falling demand. As prices for fossil fuels decrease, more demand is created. It should be noted that this is only true before the realities of peak fossil energy manifest themselves.[107] Once we are well into the downside slope after the peak, efficiency gains will not be enough to substantially lower the cost of fossil energy - but, at this point, we will no longer have 'extra' fossil energy to invest in energy efficiency technologies like wind and solar.

The Jevons Paradox is not completely unrelated to the illusion of dematerializing the economy. In the face of the many economic contradictions in the functioning of capitalism and the environmental realities of exploiting a finite earth, some people believe that capitalist economies can be 'dematerialized' to sustain growth without expanding the need of energy and resources. Some of this can be (and has been) the result of running out of crap to buy in developed nations -

people go to the theatre instead of buying a fourth sea-doo. But it doesn't take a lot of imagination to visualize that this sort of dematerialization is limited in an economy focused on planned and perceived obsolescence. Some of the gains in dematerializing the economy, it is hoped, will occur with greater efficiency gains from new technologies - though this is refuted by The Jevons Paradox. And some gains in dematerialization can simply be assigned to off-shoring energy and pollution-intensive industries to developing nations. This is precisely argued by Minqi Li: "Claims of the advanced capitalist economies to dematerialization in the wider, more meaningful sense of declining overall environmental impact are in fact refuted by the Jevons Paradox, which says that the increased efficiency in the throughput of energy and materials normally leads to an increase in the scale of operations, thereby enlarging the overall ecological footprint. ... Moreover, part of what is referred to as dematerialization arises from the relocation of industrial capital from the advanced capitalist countries to the periphery in pursuit of cheap labor and low environmental standards."[108]

The implications of The Jevons Paradox (or Rebound Effect) are clear in the case of mitigating climate change. The second labor of Hercules was to slay the Lernean Hydra - a multi-headed serpent that terrorized the countryside. When they confronted each other Hercules grasped the serpent and the serpent grasped Hercules in a dual to the death. When Hercules bashed one of the Hydra's nine heads, two would grow back. (That's the point of the story, by the way - slay one problem and two grow back ...). To finish the story, Hercules got some help from a buddy who burned off the heads of the Hydra as they tried to grow back from the decapitated stump. Let's just say that we don't have any stump-cauterizing buddies right now. A recent study by Jenkins, Nordhaus and

Shellenberger has evaluated the implications of this rebound effect for climate mitigation policy. They summarize "how multiple rebound effects operate at various scales, and describe rebound as an 'emergent property' with the greatest magnitude at the macroeconomic, global scale relevant to climate change mitigation efforts. Rebound effects are real and significant, and combine to drive a total, economy-wide rebound in energy demand with the potential to erode much (and in some cases all) of the reductions in energy consumption expected to arise from below-cost efficiency improvements."[109]

The Jevons Paradox raises significant concerns about the efficacy of efficiency-giving technologies in reducing total energy use and reducing carbon emissions. It challenges the expectation that greater efficiencies offered by alternative technologies will actually result in the decarbonization of the economy, or that it will actually allow us to match declines in the availability of fossil energy after peaking. This is particularly true since the highest rebound effect is expected in transportation, and space heating & cooling - exactly the areas where there is the highest use of fossil energy. This is why Jenkins, Nordhaus, and Shellenberger call it 'backfire'.

"Economists, however, have long observed that increasing the efficient production and consumption of energy drives a rebound in demand for energy and energy services, potentially resulting in greater, not less, consumption of energy. Energy productivity improvements over time reduce the implicit price and grow the supply of energy services, driving economic growth and resulting in firms and consumers finding new uses for energy (e.g., substitution). This is known in the energy economics literature as energy demand 'rebound' or, when rebound is greater than the initial energy savings, as 'backfire'."[110]

Jess Jenkins, Ted Nordhaus, and Michael Shellenberger

The Economist-Who-Shall-Not-Be-Named

When you ain't got nothing,
you got nothing to lose

Bob Dylan

The proletarians have nothing to lose but their chains.

Not Bob Dylan

One of the aspects of the 'I = PLOT' relationship that is rarely spoken of is the context within which our population, our lifestyle, and our technologies operate - that is, our system of Organization. And the system of organization is premised on the mode of production - our economy. Or, more precisely, our global capitalist economy.

As Žižek has said, there are known knowns - things we know we know; there are known unknowns - things we know we don't know; and there unknown unknowns - things we don't even know we don't know. Clearly, the latter is inferred from the fact that we couldn't have known we didn't know something until we became aware of not knowing it. What is rarely discussed (even by sages like Donald Rumsfeld) in the pairing of knowing and not knowing, are the unknown knowns. Žižek has called this ideology - those things we think we know and act as if we know, but remain largely unexplored and unknown. Capitalism has become such an ideology.

Everyone is familiar with the ideology - "Capitalism offers people the freedom to choose where to work and what they do." [Bush]; "Capitalism demands the best of every man - his rationality - and rewards him accordingly. It leaves every man free to choose the work he likes, to specialize in it, to trade his product for the products of others, and to go as far on the road of achievement as his ability and ambition will carry him." [Rand]; "History suggests that capitalism is a necessary condition for political freedom." [Friedman]; "Many of those who attack capitalism know very well that their situation under any other economic system would be less favorable." [von Mises]; "Capitalism is based on self-interest and self-esteem; it holds integrity and trustworthiness as cardinal virtues and makes them pay off in the marketplace, thus demanding that men survive by means of virtue, not vices." [Greenspan]. Capitalism is virtue, and it is freedom: freedom is good, so capitalism is virtuous.

There are words that have meaning, and there are words that are empty vessels that have to be filled by the speaker and hearer - virtue and freedom are empty vessels, and have been the focus of human thought (and human conflict) for a few millennia or more. Empty words are ideological words - we think we know what they mean but a roomful of people would come to blows before agreeing on a definition. Empty words are ideological words - the people who have the power get to fill the empty vessel with content - the hegemonic moment. Empty words are ideological words - they are unknown knowns.

Other than lung-bracing freedom and blue-sky virtue, what is capitalism? As the current global mode of production, how does it determine what we do and how we do it? The Economist-Who-Shall-Not-Be-Named spent a good part of his life looking behind the appearances of capital in order to understand the dynamics of the system. The following will be a brief overview of the system of capital, its contradictions, and how it is designed to eat our civilization alive.

The appearances of capitalism involve the prices of products as determined by supply & demand in the marketplace. Profits are made from the 'risk' of investing money in capital to participate in the competitive production and marketing of a commodity or service. People are free to sell their labour to whomever they like and share the revenues of the company by receiving their fair share as wages, determined by his or her ambition and the magic of the invisible hand in the marketplace. Clearly, the virtuous are rewarded, and the indolent are punished in such a natural system.

What did the Economist-Who-Shall-Not-Be-Named find when he looked behind the curtain? He found a cranky old man working a couple of levers in an effort to control the Land of Oz. So, let's start there - how did the cranky old man get behind the curtain? Ideology doesn't ask this question: where did the capitalist get the money to buy capital that can be 'risked' to produce commodities for the marketplace? The tale told is that these individuals were smarter, and saw opportunities where others did not; they were willing to take risks, where others recoiled in fear; they abstained from luxury and waste and embraced the Puritan work ethic, while others wallowed in self-indulgence. And, when, on occasion, someone actually did amass wealth by some combination of vision, ability, opportunity, and guts (with a dash of luck) they were beatified in the public forum. That's the story, but the reality is quite different.

One of the myths that conceal this reality is that a family believes they are wealthy, that they have achieved success, when there is a roof over their head, a car parked on the street, and enough savings to live a short while without an income. Most people will work their whole lives trying to achieve just this. It is no wonder that a study on financial fragility[111], when asked if they could '*come up with*' $2000 in 30 days, discovered that almost 50% of adult Americans indicated that they could not. And by 'come up with' they meant getting $2000 in their hand in any possible way - from savings, from relatives, by working extra hours, by borrowing, through credit cards ... anything. In other words, people across a wide range of incomes and wealth are barely managing to make ends meet. Some of these people have recklessly overextended themselves and it may be difficult to concede too much sympathy to them, and the survey results might also indicate a pessimistic attitude towards their society rather than an actual inability to come up with the money. This may be the case in a minority of situations - but other data supports the conclusion that most people in the United States are living hand to mouth with no reserves.

One certainty is that a majority of people in the United States are very vulnerable to financial shocks (i.e., unexpected expenses or loss of a job). Britain and Germany show similar tendencies, while countries with greater levels of welfare have roughly half the rates of financial fragility. The mechanisms the authors say that might give rise to this situation include disincentives for poorer people to save, gambling on credit, weakening social ties and lack of financial knowledge. What isn't mentioned (Occam might be disappointed), is that a large number of people are, quite simply, barely able to sustain themselves due to a combination of low wages and high costs of living - more about that, later.

Where's the money? According to a 2010 study by Norton & Ariely[112], most people have no idea what the distribution of wealth is in the United States. In general terms, the top 1% of U.S. households own 35% of the wealth and the top 20% of households own 85%. The other 80% of households own 15% of wealth (of which, the bottom 40% have practically nothing). It is interesting to note, when one considers the recent bailouts to financial corporations and the corporate lobbying-power in government, that stock ownership follows a similar distribution. The median family income (where half of the population earns more and half earns less) is approaching $50,000. The median net worth is $140,000 of which $100,000 is equity embedded in the home. We are mainly speaking here of Caucasian households, as families representing other races, when disaggregated, have much, much less.

Statistics like this are a little depressing, but that hasn't stopped you reading so far ... Income (or the money paid for your 'work') has a distribution similar to that of wealth. The average income of the bottom 90% of people in the United States is approximately $30,000. This is considerably less than the median income of $50,000 cited above, which shows just how much the top 10% earns to create such a large shift in the average. In the last 40 years, the income of the bottom 90% has dropped 1% in its purchasing power - in other words, wage increases have not kept up to the cost of living which has steadily increased over the same period of time. The top 5 to 10% interval of the population has an average income of $130,000, the top 1 to 5% interval earns over $200,000, the top 0.5 to 1% earns $450,000 on average, the top 0.1 to 0.5% earns almost $900,000 each year on average, and the top 0.1% earns (for lack of a better word) a whopping $5.6 million dollars on average. Interestingly, over the past 40 years the income of this top group has increased 385% in purchasing power. Most of the people making this sort of income are executives and financial managers who now earn over 400 times more than the average salary in the companies they manage. These managers earn in a single day what their average employee does in a year. Yes ... this top 0.1% must be very smart, innovative, and virtuous to earn that sort of coin.

"The Roman historian Plutarch, whose histories were well known to many of our Founders, had warned that an 'imbalance between rich and poor is the oldest and most fatal ailment of all Republics.'"[113]

Al Gore

The recent Occupy movement was derided in the corporate media as being incoherent: the common refrain was, "what do they want?" That's like asking what the canary wanted when it dropped dead in the coal mine. The Occupy movement is an indicator - the Republic is ailing. An interesting study validates the concerns of the foot-stomping, park-occupying precariat: In the last thirty years, a period in which the economy has grown leaps and bounds, the percent growth in household income was 390% for the top 0.1% of households, 224% for the top 1% of households, and a limp 5% for the bottom 90% of households[114] - and even that amount of largesse seems to be anathema for the wealthy as they continue their assault on the working class. The top 1% of households has captured almost 60% of all the income growth over this period, and this gap has accelerated over the past decade.

Before we move on, keep in mind we are talking about one in a thousand families earning these enormous sums, and it is about one in five families that control most of the wealth in the United States. Other nations will differ slightly, as measured by the GINI coefficient, but this type of distribution of wealth and income is more common than one would think. It is also important to keep in mind that the bottom 50% of households have very little - that is, every second family live on precarious incomes and no wealth. Clearly, this is because they are encumbered by disincentives to earn, or are simply lacking in virtue.

The question we asked, before we considered what a wealthy person might earn and own, was: where does the wealth come from to invest in capital? For the very rich today, the wealth comes mainly from owning capital - the revenue from 'risking' their money in the marketplace. But this is a tautology ... from where did the wealth originally come from? To answer that, one would have to ask what creates wealth.

For the past couple of hundred years, since Adam Smith and Ricardo, economists have largely agreed that wealth is derived from human labour and the 'free gifts' of nature. It is labour that transforms a natural material or harnesses a natural process or even converts knowledge into a commodity (a product or service). There are some forms of wealth that do not fit well into this labour-theory, such as the wealth of land ownership or the wealth of social solidarity and cooperation ('social capital'), but even these could be considered values that are dependent on an economy of labour - for what is the value of land if there are no buildings being built, no agriculture, or no resources to be extracted; and what is the value of social capital if it is not made manifest in the creation of wealth?

Wealth is derived from human labour, whether it is the original wealth used to buy capital (embodied labour), or the wealth derived from hiring labour power to use that capital. Without human labour, machines stand still, fields go fallow, and people sit around in the town square wishing they could get a haircut. The acrobatics that many modern economists perform to disguise this simple fact must surely have a motive - maybe the dream of money creating money, or wealth magically materializing from simply knowing something other people don't, or simply controlling the knowledge by making it a form of property. But, still, why would these economists put so much effort into 'discrediting' the labour-theory of value?

The Economist-Who-Shall-Not-Be-Named who took Ricardo's common-sense theory of labour-value showed why modern economists might want to veil this primary source of generating wealth. He showed that in the production of a commodity, fixed capital (including the means of production and the objects of production) and variable capital (labour-power) are used. The commodity is then sold for money, which in turn, is used to replace the fixed capital used in production and pay the worker their wages. Of course, the person with the original wealth would not 'risk' this investment without getting something for it - they want a profit. The Economist-Who-Shall-Not-Be-Named likens this impulse to the miser, the difference being that the miser increases their wealth by hording money, and the capitalist increases their wealth by keeping money in circulation, from money to commodity (capital) back to money plus an additional amount. The motive of both the miser and the capitalist, however, is to accumulate money for no other reason but to accumulate money.

Where does the additional amount earned by selling the commodity come from? If we accept the labour-theory of value, it comes from the variable capital. The fixed capital is simply replaced by the money received from selling the commodity - the cost of the machine used up in making the commodity and the cost of the raw materials used in production are recovered so as to buy them anew for the next commodity. Fixed capital (when purchased by the person with the original wealth) is a commodity - it is not human labour and it does not create value (though, certainly, they are part of the value-creating process). The additional value received when selling the commodity clearly comes from the variable capital. But, is not the person laboring paid for their labour-power at market-value? And is the money received for selling the commodity not used to buy more labour-power to make the next commodity? The Economist-Who-Shall-Not-Be-Named discerned that this is simply the surface appearance of supply & demand and the fetish of the commodity.

The labourer sells his or her labour-power in the marketplace as a commodity, which is purchased by the person with the original wealth and the fixed capital. The value of the labour-power is determined by the cost of sustaining and reproducing that labour, which will vary from place to place and time to time depending on a number of factors including the cost of living (expectations), the competition for work (unemployment), educational level, the general health of the population, and so on. The labour-power is purchased for the day and the labourer has agreed that the product of his labour is the property of the capitalist who controls the original wealth. The capitalist now employs the labour for as many hours and as intensely as possible. The profit comes from the difference between the cost of the labour (the value of the products the labourer and his or her family need to exist under the given social conditions) and the value added to the commodity by the labourer. The replacement-value of the labour-power depends on the cost of living, and the surplus-value of the labour depends on the intensity and hours of labour. The difference is translated into profit, which is reinvested in expanding production: accumulation for accumulation sake.

The Economist-Who-Shall-Not-Be-Named, to summarize, recognized that the profits (and wealth) controlled by the people with the original capital originates from the exploitation of the worker. The surplus-value can be expanded by increasing the intensity of work or the length of the workday (absolute surplus-value). The history of labour has been an effort to moderate this exploitation. The surplus-value can also be expanded by reducing the cost of reproducing the workforce, by reducing the cost of the commodities required for sustenance (relative surplus-value).

It is not difficult to see how this plays out in our modern economy. As the cost of labour increases (due to societal expectations or lack of supply), the people with wealth capitalize - they replace the labour with labour-saving machines and processes. Fewer people are required to produce the commodity, and the discarded laborers fill the pool of unemployed, putting a downward pressure on labour cost. The capitalization also reduces the value (and, in a competitive industry, the price) of commodities and reduces the cost of living. The wage paid to labour (though reduced) is still adequate to sustain the workforce as the products necessary for their sustenance become cheaper. Capitalists may also look to increase their surplus by using cheaper labour in regions that have lower costs to sustain the worker because these workers have lower expectations. This has the effect of reducing the wages and the expectations in 'uncompetitive' regions - in other words, globalization is used to 'discipline' the workforces of more developed regions. On the other hand, as capitalists continue to accumulate profits (by exploiting the worker) there is an expanding demand for labour which puts an upward pressure on wages, which must again be tempered by capitalization and/or finding lower-cost labour elsewhere.

The important point to be made is that the system requires poverty to discipline labour; the system requires that wages are maintained at the level of sustenance for the workforce; and the system requires expanding capitalization of production to absorb the surplus (by creating new markets, new products within the market, or investment in more labour-saving devices). The fact that poverty has never been eradicated or full employment realized, even in the golden age of capitalism, is not a mystery. And that half the population of the United

States is living hand-to-mouth, and that 85% of the wealth is concentrated in the top 10% of the population, is also not a mystery. This is how the system must operate. Finally, expanding capitalization, globalization, and consumerism illustrate the modern effort to reinvest the surplus. All of this is derived from the realization that profits come from separating workers from the means of production and making them sell their labour-power as a commodity for subsistence wages.

This realization, however, is disguised by the surface appearances of profits being realized by the money 'risked' by the people controlling wealth and the vagaries of supply & demand. Money makes money, and the social relations between workers and owners/executives (the modern 'personifications of capital') are veiled. The worker remains 'free' to choose to work and where to work in the marketplace of labour, and the worker is paid what he or she is worth in this marketplace. In the Land of Opportunity it is only initiative and virtue that holds a person back from untold wealth. It is no wonder why modern economists wish to discredit the labour-theory of money - it makes the wealthy look like miserly exploiters. (And we certainly wouldn't want to do that!)

We still haven't discovered, however, where the original wealth came from - all of what we have been discussing unfolds once this original wealth has been accumulated.

Given that wealth is derived from human labour, it is not difficult to discover the origin of accumulated wealth. One must simply find the people who had (and still have) the power to expropriate the value of labour-power. Accumulated wealth originated with the aristocracy of feudal times, slave owners, people with the power of taxation, and the people who control the apparatus of violence - anyone, in fact, who could force people from a life of subsistence on common lands into a life of selling their labour-power for their subsistence, and anyone who could dispossess people of their property. This has been so since the enclosing of the commons in Britain, and this 'accumulation by dispossession' continues today on a worldwide scale.[115]

Consider China today. China has roughly 20% of the global population living on 9% of the world's arable land and using 6% of its fresh water. Until recently, they have been able to provide the basic necessities for themselves, having one of the most equal distributions of wealth on the planet and a 6000 year old civilization to ground it. In what has been called 'The Great Reversal', China has become a modern day example of 'accumulation by dispossession' as the basic support for the mass of the population (the 'iron rice bowl' in the cities, and the 'clay rice bowl' in the rural areas) has been wrestled away from the people, while a distribution of wealth not dissimilar to the United States has emerged. Studies show that there are more than 250,000 millionaires in China today, holding 70% of the wealth. [116] Of the wealthiest 20,000 people, over 90% are related to senior government and Communist Party officials.[117] In comparison to the United States, you probably won't be surprised to learn that two-thirds of U.S. Senators and over half of the Congressmen are millionaires.

The capitalist system must continue to accumulate. Without surplus generation (profits) from investing capital, such investment would stop - unless, of course, these holders of wealth decide to invest their money to produce commodities for the good of society and without regard to profit. To think … a zero-growth economy in which there is no more accumulation, no need for credit, no inflation, no exploitation … but, alas, this doesn't sound like capitalism any more. Because capitalism must continue to accumulate and grow, it must continue to expand the magnitude of the production of commodities (the market), and it must increase the ability to consume the products (though population growth or consumption-oriented lifestyles).

Are there barriers to continuous expansion? The Economist-Who-Shall-Not-Be-Named thought so, as do many others outside of orthodox, neoliberal economics. The first, and most obvious, is that you need consumers to buy the commodities (goods and services) being provided by the personifications of capital. The same worker who is continually being squeezed to lower expectations and lower wages for an ever more precarious job, is the same person who is being asked to act as a consumer of the produced commodities. It is estimated that 70% of the current GDP in the United States is related directly to consumer spending (rising from a rate of 63% between the years 1960 to 1980). The end game is an automated industry without workers producing commodities that no one has money to buy. (Of course, since there is no new value being added by labour-power, this is ultimately a steady state economy). But this isn't likely to occur: as stated, the people with the wealth aren't going to build an automated factory that doesn't produce value - assuming that the factory is built in a competitive sector of the economy.

The consumer must be able to buy the commodities produced, or the surplus value (profits) cannot be realized in the conversion of the commodity back into money. To sell commodities in this environment, wages must remain sufficient, the cost of commodities must decline, or the money has to be made available on credit. For the past four decades, wages have remained roughly equal, and though there has been modest inflation, the cost of consumer products has been declining due to the exploitation of other nations where the people must work for less (due to lower costs of living and much lower expectations). The most important factor has been the recent trend to extend credit to anyone with a heartbeat (and there are likely cases that even a heartbeat has not been a requirement).

The credit available to the working class has been nothing short of remarkable in the past few decades. The household debt compared to income has steadily grown in the United States. It was roughly 35% in 1950, meaning that if a household had an annual income of $10,000, they would owe $3500. By 1960, the debt to income ratio was 58%, and by 1980 it was 69%. By 2009, this ratio was almost 140% in both Canada and the United States.

The statistics often used to illustrate debt and income is the 'household debt service ratio' or the 'financial obligations ratio' which is similar to the former but includes some other types of debts (like automobile leasing, rental payments, household insurance and property taxes). These ratios compare the cost of servicing the debts with a household's disposable income. These also show a steady increase in household debt, though at a less alarming rate. What they conceal is the very low interest rates that have attracted people to borrow the money in the first place. In other words, households can carry more debt (compared to their income) if it is cheaper to do so. Does low interest encourage borrowing and spending to keep the economy afloat, or does a 'healthy' economy supported by strong consumer demand keep interest rates low? Or do these stimuli act as a virtuous circle … while it lasts? There are those who suggest these debt ratios are perfectly sustainable, and that consumers could quickly pay off their debts (presumably by saving). What makes these assertions difficult to believe, if we recall, is that a majority of households are too 'financially fragile' to be able to save – half couldn't come up with $2000 in a month; and even if they could, it would reduce the demand for commodities and, therefore, effect the realization of profits for the wealth-holders. No profits – no capitalist economy.

"U.S. economic growth since 2001 has been led by the expansion of household consumption, which now accounts for over 70 percent of GDP. As the majority of households suffer from falling or stagnant real incomes, the expansion of consumption has been financed by the explosive growth of household debt. U.S. household debt soared from about 90 percent of personal sector disposable income to 103 percent in 2000, and to 140 percent in 2006. By 2007, the household debt services (interest and principal payments on debt) had risen to 14 percent of disposable income, the highest on record. In the meantime, the household saving rate (the ratio of household saving relative to disposable income) has fallen from the historical average of near 10 percent to virtually zero now."[118]

Minqi Li

This discussion was intended to show that one of the mechanisms of keeping the money/commodity/money-plus-profit cycle going for the owners of wealth is the extension of credit to keep people buying the commodities. A lack of consumer 'confidence' is a palpable fear on Wall Street. When consumers stop buying, the production of commodities slows, which leads to 'rationalizing' the workforce and a reduction of business investment (due to unrealized profits). Laying-off workers exacerbates this vicious cycle until consumer confidence can be restored. During the great depression, during the Keynesian revolution, stimulus was concentrated on the worker - creating jobs, and creating demand (and the world war that followed didn't hurt as a form of 'constructive destruction'). During the new great depression, beginning in 2008, stimulus was concentrated on the already-rich and their corporations. This neoliberal supply-side stimulus doesn't seem to be having much effect - hmmm, I wonder why.

Instead, the already-wealthy have used these handsome handouts to inaugurate a renewed effort of 'accumulation by dispossession' and monopolization. Choosing to follow the alternative demand-side approach to stimulate the economy, however, may not work like it did in the Great Depression because the state-supported jobs that could reinvigorate production will only be created in the place where the manufacturing actually occurs (not the United States as it did in the 1930s, but in China and other low-wage regions). It appears that the off-shoring of manufacturing in the effort of capital to expand profits has disconnected the consumer from the producer in an unfortunate way, as has been made evident during these times of cooling consumption.

There has also been a concerted effort to break unions in the Western economies so as to maintain profits by minimizing wage increases and reducing 'expectations' for a higher standard of living which might include health care, opportunities for education, a robust infrastructure, and so on. Listen to what Stephen Harper, Prime Minister of Canada, said in 2011 at the World Economic Forum in Davos: "Is it a coincidence that as the veil falls on the financial crisis, it reveals beneath it not just too much bank debt, but too much sovereign debt, *too much general willingness to have standards and benefits beyond our ability or even willingness to pay for them*?"[119] – it is clear that the 'standards and benefits' he is suggesting to be curbed are not the 'standards and benefits' of his corporate benefactors.

Another telling indicator is union membership which, in the United States, has dropped precipitously since the 1980s from almost 30% of workers to less than 15%[120]. Unions were once successful, by and large, at directing a portion of the increased profits generated by greater productivity to the worker, and union jobs even floated non-union boats for a while. As improvements to productivity have flattened in western economies, so have the gains to the worker.

Though capital has proven to be resilient in the past, it is difficult to believe that consumers can borrow more without further destabilizing the system: They can't pay back their grossly extended debt (not to mention government debt), and they can't tolerate financial shock or sudden rises in borrowing rates. The Western worker is steadily losing their manufacturing and knowledge jobs to other 'emerging economies'. The Western consumer, it is clear, has played its role and is now being aggressively ushered off-stage to stand with the rest of the world's 'surplus population'.

If you can't get your workers to buy more crap that they don't need by relentlessly advertising and extending them credit, it's time to find new consumers. The best place to find them is in the same place where you found the workers who would work for less than the Western workers that were just a moment ago sent roughly packing. New workers, new consumers, new profits. These are the 'emerging economies', the 'Mexican miracle', the 'Asian Tigers', the BRIC (Brazil, Russia, India and China), and you can add an 'S' for South Africa, if you want. The markets must be expanded to absorb the commodities that are produced by an ever-accumulating capital economy. The surplus profits from each cycle of exchange must be reinvested, and there must be a desired commodity or an ever-new product to absorb the new investments in production. Planned obsolescence, changing fashion (perceived obsolescence), and waste in packaging or wars – all of these are part of the process of reinvesting surplus to increase commodity production to be purchased by an increasing population of workers who are exploited as they add value to the production, which leads to ever greater profits. Sustainable? Not on a finite earth.

Each individual capitalist is driven by maximizing their profits at the expense of the workers. The unit labour costs in the U.S. (the amount spent on labour for each good or service produced) fell 4.7% in 2009, the largest drop ever recorded[121]. They count on the workers of other capitalists to earn enough to purchase their commodities - it is, of course, ridiculous to expect this, given that each capitalist is motived by the same forces. Using the financial crisis of 2008 to secure greater profits is a long-established mechanism, as clearly outlined by Naomi Klein's *The Shock Doctrine*:

> The truth is at once less sinister and more dangerous. An economic system that requires constant growth, while bucking almost all serious attempts at environmental regulation, generates a steady stream of disasters all on its own, whether military, ecological or financial. The appetite for easy, short-term profits offered by purely speculative investment has turned the stock, currency and real estate markets into crisis-creation machines, ... Our common addiction to dirty, non-renewable energy sources keeps other kinds of emergencies coming ...[122]

If you want to bust unions, reduce wages, eliminate social security for the poor, send more of the 'surplus population' to prisons, privatize public services or the commons ... just wait for (or, indeed, orchestrate) a crisis - financial, military, ecological ... no matter. Governments have been the best buddy of the capitalist economy while they have restrained the greed of one to sustain the greed of the many.

> The state becomes involved, *inter alia*, when it comes to immigration and labour laws (minimum wages, hours

> of work and regulation of the conditions of labour), the provision of social infrastructures (such as education, training and health care) that affect the qualities of labour supply and policies designed to maintain the reserve army (social welfare provision).[123]

But these same capitalists, through the media empires they own, have made every effort to convince the working population that the government is inefficient, wasteful, and interferes too much with the freedom of the market. Ironically, the government that restrains excessive exploitation through restrictions of the working day, child labour, minimum wage, worker's freedom to unionize and protections from predatory lending and environmental pollution, is the same government being reviled by the people who most benefit from the system. It seems that they want it both ways - a big government to absorb military production; a big government to provide the military and diplomatic might to enter intransigent markets; a big government to incarcerate the surplus population; a big government to provide research; a big government to clean up the messes made once the capitalist has moved on to better things; but at the same time they want to reduce government that wastes money on keeping the surplus population fed, healthy, and educated; to reduce government expenditures on non-'innovative' science (that doesn't result in a commodity that may be privatized, controlled and sold); and so on. Just read the paper on any given day to see the Market assault on government.

Could it be that capitalism has met its long and productive end? Declining rates of resource production, environmental stresses, a reduction in per capita agricultural yields, the limits of globalization to exploit workers, the limits of what the people can afford to consume, and a dog that bites the government hand that feeds them - are these the indicators? And as corporate demands from labour-power intensify, people are receiving fewer benefits for their work.

> At the other end of the wealth scale, those thoroughly incorporated within the inexorable logic of the market and its demands find that there is little time or space in which to explore emancipatory potentialities outside what is marketed as 'creative' adventure, leisure, and spectacle. Obliged to live as appendages of the market and of capital accumulation rather than as expressive beings, the realm of freedom shrinks before the awful logic and the hollow intensity of market involvements.[124]

Given that this is our global Organization - a system in which the few exploit the many in an ever growing cycle of production and consumption, a system in which we are currently dependent for our sustenance - it is difficult to see how it can be changed. That this type of Organization is vulnerable to its own internal contradictions is obvious to those who can see, but what is less obvious is how a steady-state economy that satisfies our collective needs will emerge. It is unlikely, if it does emerge, that it will do so without a lot of suffering, as those with the power try to maintain the existing system. Rosa Luxemburg once presented the option as being either an economy premised on democracy and the satisfaction of needs, or barbarism. How does barbarism look to you?

More on Organization

So foul a sky clears not without a storm.

Shakespeare

"Nations around the world, and especially the poorest countries and communities, confront a multitude of interlinked challenges and pressures. These include rising competition for resources, environmental breakdown and the specter of severe climate disruptions, a resurgence of infectious diseases, demographic pressures, poverty and growing wealth disparities, and convulsive economic transformations that often translate into joblessness and livelihood insecurity. ... A number of these conditions and dynamics can be seen as an outgrowth of the dominant economic model premised on essentially unlimited resource consumption. This model is not only putting humanity on a collision course with the planet's ecological limits, it has also lead to tremendous social and economic inequality."[125]

Michael Renner

This system of Organization is no longer providing for the people it is meant to serve, and is destroying the foundations of wealth by consuming and destroying the natural environment. Leonard expresses this succinctly:

> Almost every indicator we can find to measure our progress as a society shows that despite continued economic growth over the past several decades, things have gotten worse for us. In the United States, obesity is at record levels, with fully a third of adults over the age of twenty and nearly 20 percent of children between the ages of six and eleven considered obese. A 2007 report revealed a 15 percent rise in teen suicides between 2003 and 2004, the largest increase over a single year's time in fifteen years. In 2005 we had ten times as much clinical depression as in 1945. The use of antidepressants tripled between 1994 and 2004. As many as 40 million Americans are now allergic to their own homes – to the chemicals in paints, cleaning products, treated wood, wallpaper, and plastics. We average 20 percent less sleep at night than we got in 1900. Americans work more hours than people in almost any other industrialized country. The debts of individual consumers have been growing at twice the rate of incomes. ... Despite the spending beyond our means, our country still faces devastating levels of income inequity, poverty, homelessness, hunger, and the lack of health insurance.[126]

The market, in the words of Galbraith, "has its own truth on which reality does not intrude"[127] - and that truth is limitless growth. In the process of atomizing communities into consuming individuals, this form of Organization has transformed what we value, and it is an understatement to say the "values of a society totally preoccupied with making money are not altogether reassuring."[128]

By fixing men's minds, not upon the discharge of social obligations, which restricts their energy, because it defines the goal to which it should be directed, but upon the exercise of the right to pursue their own self-interest, it offers unlimited scope for the acquisition of riches, and therefore gives free play to one of the most powerful human instincts. To the strong it promises unfettered freedom for the exercise of their strength; to the weak the hope that they too one day may be strong. Before the eyes of both it suspends a golden prize, which not all can attain, but for which each may strive, the enchanting vision of infinite expansion. It assures men that there are no ends other than their ends, no law other than their desires, no limit other than that which they think advisable. Thus it makes the individual the center of his own universe, and dissolves moral principles into a choice of expediencies. And it immensely simplifies the problems of social life in complex communities. For it relieves them of the necessity of discriminating between different types of economic activity and different sources of wealth, between enterprise and avarice, energy and unscrupulous greed, property which is legitimate and property which is theft, the just enjoyment of the fruits of labor and the idle parasitism of birth or fortune, because it treats all economic activities as standing upon the same level, and suggests that excess or defect, waste or superfluity, require no conscious effort of the social will to avert them, but are corrected almost automatically by the mechanical play of economic forces.[129]

In addition to the atomization of the community into an assortment of *consumatons*, the logic of endless growth upon which capitalism is premised is untenable - particularly when this growth is derived from waste and destruction: "What is at stake is not really capital's *efficiency*, which might be imposed by a more or less drastic reallocation of economic resources, but, on the contrary, the very nature of its *productivity*: a productivity that necessarily defines itself through the imperative of its relentless, alienated self-expansion as a *destructive productivity*, which unceremoniously demolishes everything that happens to stand in its way."[130]

Flavin & Gardiner argue that global ecosystems and resources "are simply not sufficient to sustain the current economies of the industrial West and at the same time bring more than 2 billion people into the global middle class through the same resource-intensive development model pioneered by North America and Europe. ... In the coming decades, we will either find ways of meeting human needs based on new technologies, policies, and cultural values, or the global economy will begin to collapse."[131] I have already argued that new technologies are not going to make (the more necessary resource) water from wine, and it will be argued that policies are written to reflect the needs of the hegemons and that cultural values are largely inelastic (especially without 'universal' or utopian meaning or goals that can resonate across the contingent particularities of our society).

Particularly discouraging is the desire to fix the problem with the same tool that caused it. This speaks of our inability to view the problem from outside of the problem - or, in other words, to see outside of our ideology. We have an economic Organization that must grow, and it will consume our resources and destroy our environment to achieve this growth, while at the same time immiserating most of the people this form of Organization is ultimately meant to serve. As an example, one the most dangerous challenges to humanity is the emission of greenhouse gases that contribute to global warming and climate change. So, to address this challenge and reduce carbon emissions, we rely on the marketing of carbon within the same system that created the need to reduce the carbon emissions to begin with. Since our economy is energy-reliant (from heating and cooling buildings, to transportation, to manufacturing, to agriculture), there is no meaningful discussion of limiting the use of energy - the dream of substantially 'dematerializing' our economy is a fantasy. We will, instead, rely on the market to direct declining amounts of energy in the most profitable industries. The predictable result is that: "Market-based solutions to climate change, such as emissions trading, have been shown to promote profits, and to facilitate economic growth and financial wealth, while increasing carbon emissions."[132] The argument is that the money accumulated by capitalists in a capitalist economy must find a place for reinvestment (as it cannot be consumed), so even if cap & trade or carbon tax schemes might in principle redirect production to lower carbon technologies or sectors, profits will be made from emission trading, which must be used to expand the economy (in wasteful and destructive ways, including creating more carbon emissions). This can be seen as a variation of The Jevons Paradox, previously discussed.

Another source of disequilibrium in this economic system of Organization is that we have privatized the source of wealth and socialized the costs of waste. From the perspective of the natural world: "Nature is the point of departure for capital but typically not a point of return. Nature is an economic tap and also a sink, but a tap that can run dry and a sink that can clog up. Nature as a tap has been more or less capitalized; nature as sink is more or less uncapitalized. The tap is typically private property; the sink is typically the commons."[133]

Another concern is that once the contradictions inherent to capitalism manifest themselves in economic stagnation, and once the working class has been immiserated to bare levels of subsistence (with the surplus population surviving on less), this system of economic Organization will, in order to realize greater profits, begin to ignore its already deplorable concern for the natural environment. O'Connor argues:

> At the risk of grossly oversimplifying a hugely complex process, the process of accumulation through crisis has two different kinds of effects on nature. Troubled industries or regions try to save money by neglecting environmental protection and cleanup ... Also, while pollution in these industries may fall because total production declines, total pollution may increase because pollution per unit of output increases. At nature's tap, capitals in economic trouble are likely to be more ecologically careless about exploration, extraction, and processing techniques; land use; and so on."[134]

It is, therefore, not hard to imagine capital destroying the two things from which its profits are created - human labour-power and nature:

> Examples of capitalist accumulation impairing or destroying capital's own conditions, hence threatening its own profits and capacity to produce and accumulate more capital, are many and varied. The warming of the atmosphere will inevitably destroy people, places, and profits, not to speak of other species of life. Acid rain

> destroys forests and lakes and buildings and profits alike. Salinization of water tables, toxic wastes, and soil erosion impair nature and profitability. The pesticide treadmill destroys profits as well as nature. … capital's destruction of established community and family life as well as the introduction of work relations that impair coping skills and create a toxic social environment generally.[135]

The obvious summary of this is: our current system of economic Organization is at odds with the motivation and direction that we require to survive. The dream was to use the stupendous technological and productive capacity facilitated by the dynamics of capitalism for our long-term good. Now we have a system out of control - a system that we don't even recognize, as it has become the air that we breathe. It is this system that establishes our expectations of Lifestyle, and it is this system that requires a growing Population of workers to make crap for a growing number of consumers (plus a little extra population to keep the costs of labour-power low). And it is this system that creates the Technologies that inaugurate new waves of consumer growth and the capitalizing of the economic system to keep the requisite number of people waiting in the industrial reserve army of workers. It is not too wild to say, I don't think, that this component of Organization must be the fulcrum for change. To expect a radical awakening and restructuring of this destructive system, however, is delusional at best. What fish would risk living outside of water?

"These features of capitalism, as they are constituted today, work together to produce an economic and political reality that is highly destructive of the environment. An unquestioning society-wide commitment to economic growth at almost any cost; enormous investment in technologies designed with little regard for the environment; powerful corporate interests whose overriding objective is to grow by generating profit, including profit from avoiding the environmental costs they create; markets that systematically fail to recognize environmental costs unless corrected by government; government that is subservient to corporate interests and the growth imperative; rampant consumerism spurred by a worshipping of novelty and by sophisticated advertising; economic activity so large in scale that its impacts alter the fundamental biophysical operations of the planet - all combine to deliver an ever-growing world economy that is undermining the planet's ability to sustain life."[136]

James Gustave Speth

A Final Word on the Economics of Renewable Energy

Keys to the kingdom don't fit no more.

Ry Cooder

There is much that has been said about the economics of renewable energy. Most successful renewable energy projects have relied on long-term subsidies and purchasing arrangements that guarantee a market for the installed capacity. As noted earlier, subsidies make it difficult to know what the real price is for a product - remember Warren Buffet's example of a neighbourhood where people buy each other's houses. It's not that we don't know what we pay for a kWh of electricity from the distributor, or the cost of natural gas to heat our homes - what we are talking about is the subsidization of the technologies that allow the energy to be sold at these levels, and this argument includes the significant subsidies to giant corporations exploiting fossil fuels.[137]

Feed-in-Tariffs (FITs) or Standard Offer Programs (SOPs) provide an incentive to invest in renewable technologies by buying the electricity at prices that help pay for the technologies within the lifetime of operation. The costs of the technologies are also artificially reduced at the supply end of the chain, where manufacturers are subsidized directly, or subsidized indirectly through tax holidays, free land, workforce training and the like.

The economics that might be interesting to explore a little further relates to the labour-theory of value (as discussed). If value is added to a product only through human labour that transforms the 'free gifts' of nature into commodities, then what adds value when implementing a solar PV module or wind turbine? Once installed, the technologies produce electricity which can then be sold - but can the total value produced by the PV module or wind turbine over its life-cycle ever exceed the value embodied in the technology? If no additional labour (variable labour) is used to employ fixed capital (embodied labour), then, according to the labour-theory of value, no additional value is created in the commodity. This might also be reflected in the shadowy appearances of prices in the market.

This is difficult to see today because of the subsidies provided for all forms of energy generation, and in our example, renewable energy technologies. These values also do not include the externalities (like the costs of GHG emissions), which renewable energy technologies are being developed to mitigate. In a simple example: let's say you purchased a solar PV module (fixed capital) and installed it on your roof. There may be a little labour power added to maintain the system, but the module will ostensibly produce energy thereafter without human intervention. How can it produce more value that was embodied in the technology? At the appearance level of prices, it would be like a money generating machine. I can purchase a solar PV module and, over time, get more money for the electricity produced than what was invested ... money for nothing. We should all get one. But we aren't all racing to the store ... why? Because it is too good to be true. Some might argue that if everyone generated electricity the prices would decrease (because of supply and demand) and the lower prices would curb investment to a more sustainable level (at the point where you can get more free money from another investment). If the PV module made free money, why would the manufacturer sell it for less than the profits that could be realized (after installation costs and land rent).The real reason is that, because there is no labour power to exploit, there is no surplus value produced once installed. No surplus value - no profit. Renewable energy without subsidies will never be an attractive investment in the free market for electricity.

This illustrates the perversity of renewable energy in a capitalist system that is designed to exploit nature and labour to achieve profits. Not only do we not have enough capacity to adequately increase the production of renewable energy technologies, and not only do these technologies need an up-front investment of fossil fuel energy, but they also add no additional value (surplus labour-power) during operation. As such, the vision of alternative energy powering a capitalist economy is chimerical.

Engineering Fantasies

Since renewable energy remains a low-EROEI substitute for fossil fuels, that there is not enough manufacturing capacity to build enough to even remotely approach the demand for energy, and that it remains (and will continue to remain) uneconomic in a capitalist organization of economics, these technologies will not likely be a solution to our environmental challenges. What, then, might actually work in a crazy system that profits from scarcity and destruction? Well, of course, even crazier technologies that can create even more destruction! Welcome to the world of 'geo-engineering'.

There are two main approaches to geoengineering the climate: the first is to remove carbon dioxide from the environment, and the second is to modify the albedo (reflectivity) of the earth - the more of the sun's radiation that is reflected, the cooler will be the planet. Some of the most popular ideas include spraying sulfur into the air to reflect the sun's radiation before it enters the atmosphere; or spraying ocean water to create reflective cloud cover; or, fertilizing the oceans with iron filings to promote algal growth and absorb atmospheric carbon before sinking to the bottom; or, get this, putting a great big umbrella in space to shade the earth. Hansen (2009) cites several reasons why geo-engineering might be a bad idea:

> First, carbon dioxide must be less than 350 ppm to avoid ocean acidification problems. Second, sun shielding at present is far more expensive and difficult to implement than rational alternatives such as energy efficiency, renewable energy, and nuclear power. Third, it is generally a bad idea to try to cover up one pollution

> effect by introducing another; such an approach is likely to have many unintended effects.[138]

Hansen does not mention the moral hazard of geoengineering, such as failing to deter the combustion of fossil fuels with the belief that the impacts can be mitigated in another way or in the future: "Other criticism comes from those who see geo-engineering projects as reacting to the symptoms of global warming rather than addressing the real causes of climate change. Because geo-engineering is a form of controlling the risks associated with global warming, it leads to a moral hazard problem. The problem is that knowledge that geo-engineering is possible could lead to climate impacts seeming less fearsome, which could in turn lead to a weaker commitment to reducing greenhouse-gas emissions. It could be argued that pursuing geo-engineering solutions sends the message that humans can continue to live out of harmony with the Earth as long as we have enough clever technological solutions to preserve human life. This disregard to the overall health of Earth's ecosystem and natural environments is an affront to proponents of sustainable development."[139]

A recent study by The Royal Society on geoengineering the climate clearly states: "None of the methods assessed offers an immediate solution to climate change and too little is understood about their potential future effectiveness, risks and uncertainties to justify reducing present and future efforts to reduce greenhouse gas emissions."[140] The Royal Society recommends more study on life-cycle benefits, and long-term impacts of geoengineering the earth's climate. They recognize, however, that without effective action in the near term, these technologies may represent the last ditch effort to save ourselves. Of course, by this time, we will not have the energy sources to both sustain a large human population and to invest in large geoengineering schemes, and each passing year will mean less energy to divert to these schemes. Have you ever seen the movie, 'Soylent Green'?

In principle, these geoengineering schemes carry with them the potential for even greater ecological disasters - Hercules and the Hydra.[141] What is the effect of blocking the sun's radiation from the earth (even beyond Hansen's observation that the carbon will still be emitted, and oceans will continue to acidify)? What are the long-term effects of continually spraying sulfur into the atmosphere, and what happens if we have to stop for some reason? What happens within the oceans after the mother of all algal blooms? I'm surprised we haven't put more thought into just discarding this old planet and go off to look for a fresh one.

The only good thing that will come of geoengineering is that it will provide more opportunities for capital - benefiting from insecurity and scarcity. And this leads us to The Lauderdale Paradox ...

The Lauderdale Paradox

Similar to The Jevons Paradox, The Lauderdale Paradox describes the perversity of capitalism to create scarcity in an effort to create a profitable opportunity to supply what was once free and available to all - some call this tendency 'the tragedy of the commons'. Unlike the tragedy of the commons, however, this isn't the result of some perceived greed innate in the human character. This is a tragedy that is purposely manufactured in a society (Organization); it is a tragedy that actually *promotes* scarcity and greed - capitalism. This is not to say that some societies have not fouled their own nest from mismanagement and greed - but it would be difficult to show that these societies destroyed themselves in a purposeful effort to corner the market for some essential commodity.

The Lauderdale Paradox is named after James Maitland, the eighth Earl of Lauderdale (1759-1839) who wrote *An Inquiry into the Nature and Origin of Public Wealth and into the Means and Causes of its Increase* (1804). Lauderdale argued that there was an inverse correlation between public wealth and private riches, and that an increase in riches tended to diminish public wealth. "Public wealth," he wrote, "may be accurately defined ... *to consist of all that man desires, as useful or delightful to him.*" Such goods have use value and thus constitute wealth. But private riches, as opposed to wealth, require something additional (i.e., they have an additional limitation), consisting "*of all that man desires as useful or delightful to him, which exists in a degree of scarcity.*"[142]

In short, if one can create scarcity in an essential commodity and if one can control the distribution of that commodity, then one can make heaps of money. As an explicit example, look at rebel militias controlling the mining of coltan in the Congo – a scarce metal we use for electronic devices; or, closer to home, look at the military deployment in the Middle East to control the flow of scarce oil. More indirectly, since the advent of corporatist-based governments, is the disappearance of the public commons (public space, clean air and water, social practices, etc.) as it is transitioned to private control. As such, even absolutely necessary commodities like water have been privatized by corporations to be delivered to those who can pay - ignoring the basic right to life. In the late 1990s, Bolivia sold their water rights to two multinational corporations that not only threatened to prevent access to this necessity by the poorest, but prohibited these people from collecting rainwater from their roofs as they had done for centuries. Uprisings in the nation encouraged the government to reverse these agreements, but it remains a good illustration of The Lauderdale Paradox - making clean water scarce, controlling it, and then profiting from its sale.

What makes this particularly pernicious with respect to climate change and peak oil is that the negative consequences of scarcity become beautiful opportunities to profit. Declining fossil energy, scarce water, the degradation of arable land, food scarcity, and even clean air can become opportunities for control and profit. Why would anyone in their right mind try to prevent these opportunities from emerging?! Foster and Clark cite Herman Daly on The Lauderdale Paradox: "The ecological contradictions of received economics are most evident in its inability to respond to the planetary environmental crisis. This is manifested both in repeated failures to apprehend the extent of the danger facing us, and in the narrow accumulation strategies offered to solve it. The first of these can be seen in the astonishing naiveté of leading orthodox economists - even those specializing in environmental issues - arising from a distorted accounting that measures exchange values but largely excludes use values, i.e., issue of nature and public wealth."[143]

In summary, the reason 'we couldn't do anything about it if we did understand the problem' can be largely attributed to our forms of Organization - specifically our economic organization, or mode of production. This is a system that must by definition continuously expand, and expanding the economy on a finite planet that is expected to provide the resources and absorb the wastes is not possible. This economic system also has a tendency to transform efficiency into growth (rather than into the conservation of resources). This is The Jevons Paradox as it relates to the functioning of a market economy. In other words, if we try to reduce our collective impact through more efficient technologies, we will instead reduce costs and spur greater consumption. We also have The Lauderdale Paradox, which describes the potential for control and profit-making when what was once abundant and common becomes scarce. What incentives are there to avert risks or reduce pollution when they create opportunities for profit? Geoengineering, in many ways, illustrates this paradox: we will emit greenhouse gases into the atmosphere, knowing full well that the ultimate results will be disastrous, but then create a technology that can be profitably employed to try to reduce the impact of global warming. We could choose to simply reduce our emissions, but how could that be profitable? Clearly, our current form of global economic Organization is a substantial barrier to a rational response to our many environmental and social challenges. And what is worse, this economic Organization operates at the level of ideology - unknown knowns that we cannot see and therefore cannot even begin to address.

Popping Population

All God's children they all gotta die.

Nick Cave, Murder Ballads

Then there is the elephant in the room. A world population that has exceeded 7 billion and is expected to grow another 50%, to over 10 billion by the middle of the 21st century.[144] As discussed, population is one of the key elements of our collective impact but it is one that is rarely addressed by governments. Why is that? The most likely answer is that controlling population growth infringes on the rights of people to have children, and it is associated with sensitive issues like abortion and contraception which is complicated by the influence Organizational systems like culture (marriage, sexuality) and religion. The fact of the matter is that the more people there are on earth, the greater will be the pressure on deforestation so as to increase arable land to grow food, and the greater will be the extraction of resources (water, energy, and minerals) to meet minimum basic needs. Regardless, it will be argued, nothing substantial will be done to curb population growth.

As discussed earlier, the relationship between Population growth and Impact is also dependent on Lifestyle expectations, Technology and (institutional and economic) Organization (I = PLOT). So, even though the less developed regions will contribute most to the growth in population, the population in the more developed regions may continue to have more Impact per capita. Population management in the more developed regions of the world should be the focus - otherwise, lifestyle expectations must be curbed. Or, the global population may be curtailed more mercilessly by food insecurity, disease vectors, security and migration stresses, and other impacts of climate change - and those mechanisms would be more natural and absolve us of ultimate responsibility, wouldn't they?

There are arguments "that climate change will contribute to the following kinds of insecurities: tensions over scarce resources; land loss and border disputes; conflicts over energy sources; conflict promoted by migration; and tensions between those whose emissions caused climate change and those who will suffer the consequences of climate change."[145] It's hard not to think of the Pitri dish metaphor as the bacteria consume their environment and cause a precipitous crash in population.

As stated, there are cultural, religious and rights barriers to addressing population - particularly by elected governments. Another barrier is the Organizational system, as it relates to the last section on economics. It is informative to quote the economist-who-shall-no-be-named at length:

> The greater the social wealth, the functioning capital, the extent and energy of its growth, and therefore also the greater the absolute mass of the proletariat and the productivity of its labour, the greater is the industrial reserve army. ... But the greater this reserve army in proportion to the active labour-army, the greater is the mass of consolidated surplus population, whose misery is in inverse ratio to the amount of torture it has to undergo in the form of labour. ... This is the absolute general law of capitalist accumulation.[146]

In other words, to extract wealth (surplus value) wages must be kept low by maximizing exploitation. Wages are kept low by making workers more substitutable and jobs more precarious. The industrial reserve army of workers who are sitting on the bench waiting for a job opening 'disciplines' the working class to accept lower wages: potential unemployment is a very effective lever to exploit workers. If the industrial reserve army diminishes with economic expansion and innovation, capitalists must invest in the means of production to put more people out of work. As such, accumulation of wealth and capital is, at the same time, the accumulation of poverty and misery within the industrial reserve army.

André Gorz quotes Lecher that the workforce in Britain during the 1990s could be divided into three main groups: 25% belong to the 'stable core' of people trained for their occupation; 25% have stable jobs on the periphery, meaning lower-skilled administrative jobs that can easily grow or contract with demand; and the remaining half occupy unskilled and occasional jobs and represent a practically inexhaustible industrial reserve army. [147] A more recent survey published by the International Labour Office (ILO)[148] indicates that almost 15% of the global population in their productive years (25 to 54 years) is inactive, another 5% of the population is unemployed, and 45% are vulnerably employed, constituting a global reserve army of roughly 65%. Worldwide, there are some 1.5 billion people who are highly vulnerable in their jobs, and are more likely to experience conditions that violate their fundamental rights. Almost 40% of the global workforce earns less than $2 per day, and more than half of this group is considered extremely poor, earning less than $1.25 per day (purchasing power parity).[149]

Population growth has not only been an important mechanism for the reproduction and expansion of the workforce as the economy grows, it has also been an important mechanism to maintain a downward pressure on wages and expectations (maximizing the relative surplus value). The experiences of globalization over the past number of decades should eliminate any doubt about this process. Corporations have chased low wages throughout the world and have maintained a downward pressure on purchasing power in the more developed regions. Countries maintain a constant low-wage workforce by enforcement, by urbanizing the rural population, and through population growth.

There is no enthusiasm from corporations (currently exploiting an endless supply of labor in less developed regions) to curb population growth. And there is no enthusiasm for elected officials to disagree with the needs of corporations. Furthermore, the discussion on population control is entangled in discussions on rights, on racism, and on religious doctrine. As such, population as a component of impact (I = PLOT) is highly inelastic in the downward direction.

Food or Famine

How many times can a man turn his head
and pretend that he just doesn't see.

Bob Dylan

Roughly one in every six people on the planet is undernourished - that's a billion people. They wake up hungry, they spend their day trying to feed themselves, and they go to sleep hungry. We could probably do something about this, but these hungry represent the surplus population necessary to motivate those earning just enough money to feed themselves by working - accumulation of wealth at one pole means accumulating misery at the other. Though it is said that there is enough food being produced to feed the current population of the world (that undernourishment is a distribution or income problem), there are a number of factors that could be considered game-changers: energy to produce fertilizer and pesticides; energy for transportation; water to irrigate land for food production; the risks of relying on mono-cropping in industrial agriculture; the failure of GMOs to yield greater increases in production; and soil degradation and erosion.

The world has about 14 million square kilometers of arable land and permanent crop land - the land used to grow our food. Half of this land is used to grow 2.5 billion tonnes of cereals. There is another 35 million square kilometers of land that the FAO (2010 statistics) considers to be potentially cultivable, much of it being used for pastures or commons. Of the arable land currently in use for crop production, 20 percent is irrigated to grow roughly 40% of the world's food using 70%

of all water withdrawn from lakes, rivers and groundwater aquifers.[150] These irrigated crops rely on water which may become scarcer with shifting precipitation and diminishing groundwater sources.

The amount of land being brought under irrigation has slowed drastically in the past couple of decades.[151] This may be due to the better areas having already been developed, the prohibitive cost of infrastructure like dams and canals compared to current food prices, and it may partially be due to the difficulty of displacing people from marginal lands - what is marginal for a government is, by contrast, a livelihood for people. Another factor is the challenge of water scarcity. Since 2008 groundwater has been the primary source of irrigation water and most of the newly irrigated land has depended on exploiting this groundwater. Unfortunately, many of these groundwater aquifers are being depleted at rates much greater than the rate of replenishment - many are considered fossil aquifers which are effectively non-renewable. Water is also being removed from surface water sources like rivers at rates that leave little water to maintain the necessary in-stream flows to sustain the ecosystem, leading to species stress and extinction. As the impacts of climate change advance, the area of land considered water-stressed and water-scarce will increase, which will put even more pressure on maintaining even the existing land under irrigated cultivation (and, thus, maintaining the food required to feed a growing population).

Of the cereals grown in the world, roughly a third is maize, a third is wheat, and 20 percent is rice. Of the ten thousand traditional grains, many of them being more nourishing than the grains cultivated today, only 150 crops are being widely cultivated. The reliance on fewer crops reduces agricultural biodiversity and jeopardizes global food security as the impacts of climate change challenge the ability of crop species to adapt to water scarcity and average temperature increases, as well as to changing precipitation patterns that shift the distribution of water in both time and space. Furthermore, monoculture is more susceptible to pests and diseases and tends to unbalance the ecosystem upon which the agricultural system relies (for pollination and the benevolent interrelationships of species in the soil).

Grain production has increased steadily since the 'green revolution' that was prompted by the introduction of fertilizers, pumped irrigation, expanding lands under cultivation, machinery and industrial farming techniques (mono-cropping), and more recently the introduction of genetically modified plants. Yields have increased almost 150% since the 1960s. This steady increase in production for the past half-century has roughly maintained the production rate of grains at over 300 kg per capita each year, as the global population has grown. Half of this is used directly for human consumption, a third is fed to livestock, and about 5% is currently used for the production of biofuels. Pressure to increase biofuel production and meat production (which has almost quintupled worldwide over the past half-century) means that there will be less grain available for direct consumption, and will result in higher prices which, in turn, will prevent the two billion people living on less that $2 per day (purchasing power) to afford the food they need.

Despite diminishing access to new arable land for agriculture, the deterioration of the soil, and the threat of water scarcity for irrigated lands, many believe that genetically modified plants will make up the deficit in food production. Lester Brown, however, has said that the genetic modification of crop plants "have yet to produce a single variety of wheat, rice, or corn that can dramatically raise yields"[152] Gurion-Sherman in his 2009 study, *Failure to Yield,* substantiates this claim: "Bottom line: They are largely failing to do so. GE soybeans have not increased yields, and GE corn has increased yield only marginally on a crop-wide basis. Overall, corn and soybean yields have risen substantially over the last 15 years, but largely not as result of the GE traits. Most of the gains are due to traditional breeding or improvement of other agricultural practices."[153]

There has been some suggestion that GMO species have less genetic diversity and are more vulnerable to changes in the climate compared to natural species. There are also concerns of 'genetic pollution' where unintended modification of natural plant species affects ecosystem balance: a recent study states the *"potential for genetically modified (GM) crops to threaten biodiversity conservation and sustainable agriculture is substantial."*[154] Furthermore, GMO seeds are generally reliant on greater amounts of water and require an increasing use of herbicides, pesticides and fertilizers, which then result in the poisoning of water systems including the eutrophication of many rivers and productive deltas where rivers meet the sea. Natural species of food plants tend to be more robust with changing temperatures, having developed more resilience over the millennia through which they have evolved.

Due to the growing pressure to respond to demand for food, GMO seeds have become a commercial input controlled by a small number of corporations (The Lauderdale Paradox): "The top 10 seed firms control 30 percent of the global seed market; five companies control 75 percent of the global vegetable seed market."[155] This commodification of seeds through intellectual property agreements and corporate pressure, in collusion with government force, has become a barrier for many nations to become involved in GMOs, and it has become a form of slavery for those who have adopted GMO seeds – borrowing money to buy seeds and the chemicals required to grow the seeds, and then relying on highly manipulated food prices to earn enough to pay back the loans. Thousands of Indian farmers have, as a result, committed suicide out of desperation to maintain this non-virtuous cycle.

Oligopolistic markets are roughly defined as a few companies controlling over 40 percent of the market, and able to exert influences on prices due to the non-competitive nature of the market. "In the meat sectors 87 percent (81 percent by the largest three) of the beef cattle are slaughtered by the four largest firms, and 73 percent of the sheep are processed by the four largest firms. The control of hog slaughtering by the four largest firms increased from 37 percent in 1987 to 60 percent today. ... In the crop sectors, the four largest firms process from 57 percent to 76 percent of the corn, wheat, and soybeans in the United States."[156]

In the corporate food system, food is commodified and open to speculation by agri-business and the financial markets. The recent food crises have revealed the vulnerabilities of the population to food speculation, even with ample food harvests: "Food stocks are highly centralized - five corporations control 90 percent of the international grain trade, three countries produce 70 percent of the exported corn, and the thirty largest food retailers control one-third of the world grocery sales."[157] Clearly, making food or water a commodity (as opposed to a human right) makes a large percentage of the human population extremely vulnerable to the effects of scarcity - real or manufactured by corporate manipulation.

Fertile soil hosts some of the most diverse ecosystems on the planet. Plants have evolved in these ecosystems where bacteria, insects and other species help plants absorb nutrients and fight disease. Over-exploiting this soil and disrupting the ecosystem through the use of agricultural chemicals threatens the long-term productivity of the soil. Removing the crop from the fields further degrades the soil through the progressive loss of organic materials: "Long-term research studies reveal average losses of 328 pounds of organic matter per acre per year with plowing, ... Erosion from a conventionally tilled watershed has been found to be 700 times greater than from a no-till watershed."[158] In essence, the organic materials are removed from the soil and shipped to the city, where they become 'waste' rather than being returned to the land. Fertilizers are then used to return nitrogen, phosphorous and potassium to the soil, but the many minerals required for nutrition are mined and seldom replaced. As such, the nutritional value of our food has been in steady decline.[159]

Fertilizer use has steadily increased with the growing rates of grain production and is used in the production of over 50% of the world's food supply. The nitrates and phosphorous in the fertilizers, when over applied, run off to the surface water during rainfall. The algae in the water respond well to fertilizers, resulting in algal blooms. When the algae die, they sink to the bottom of the river or lake and decay, and in this process the bacteria absorb oxygen from the water (also called biochemical oxygen demand, BOD). Life in the water is greatly diminished when there is not enough oxygen – this is called eutrophication. In other words, over-fertilization kills our rivers, lakes and the richly diverse regions where rivers meet the ocean. Furthermore, as herbicides and pesticides are applied in ever greater quantities in response to the degradation of the soil, they have become less effective: "Weeds, fungi, insects and other potential pests are amazingly adaptable. Five hundred species of insects have already developed genetic resistance to pesticides, as have 150 plant diseases, 133 kinds of weeds and 70 species of fungus."[160] As they become less effective, new chemicals have to be developed, adding to the complexity and cost of agricultural practices. At a point, we will be running to stand still.

Another factor threatening our ability to feed the global population in the future is the loss of soil to erosion and the reduction of the quality of remaining soil as it is unsustainably mined of minerals. According to the FAO, 25% of the total global land area is currently highly degraded[161]: "with more than 20 percent of all cultivated areas, 30 percent of forests and 10 percent of grasslands undergoing degradation."[162] Humans currently use 40% of the biomass grown each year[163] (that is, the sustainable products of photosynthesis), leaving more soil exposed to erosion, less organic matter being returned to the soil, and less habitat for other species to use. Haberl, Erb & Krausmann suggest that the human impact on the functioning of the biosphere is considerable, and exceeds natural variability in many cases: "Up to 83% of the global terrestrial biosphere have been classified as being under direct human influence, based on geographic proxies such as human population density, settlements, roads, agriculture and the like; [and] another study estimates that about 36% of the Earth's bioproductive surface is 'entirely dominated by man'."[164]

Population pressures cause marginal lands to be brought into cultivation and deforestation. The rain forests in Brazil are being deforested at a rate of about 12,000 square kilometer each year for agricultural use[165], and the land is fertile for only a few years. Climate change exacerbates both drought and flooding which also tends to intensify soil erosion.

> Since 1945, the total land degraded by soil depletion, desertification, and the destruction of tropical rainforests comes to more than 5 billion hectares, or greater than 43 percent of the Earth's vegetated surface. … It takes 500 years for nature to replace 1 inch of

> topsoil. Approximately 3,000 years are needed for natural reformation of topsoil to the depth needed for satisfactory crop production.[166]

The three main factors contributing to soil erosion are urbanization, deforestation, and industrial agriculture. It is estimated by the FAO that 5 to 7 million hectares of land is lost to erosion each year. Worldwide, some 70 percent of the 5.2 billion hectares drylands used for agriculture are already degraded and threatened by desertification.[167] "Soils of farmlands used for growing crops are being carried away by water and wind erosion at rates between 10 and 40 times the rates of soil formation, and between 500 and 10,000 times soil erosion rates on forested land."[168] In the US, "a million acres disappear annually to urbanization and 2 million acres of farmland are lost to erosion, soil salinization, and flooding or soil saturation as a result of intensive agriculture, which consumes groundwater 160 percent faster than its replenishment rate."[169]

> As we chop down rain forests and pipe water deep into deserts, we are fast approaching the time when all land that can be used to grow food has been found, transfigured, and put into production. As we run out of places to grow food, we are pouring roads and building foundations over former farmland, strip-mining the nutrients from our soil, polluting our groundwater and soil, and filling huge tracts of land with salts so that nothing will grow there anymore."[170]

In addition to land degradation and erosion, demands to increase global biofuel production take arable land out of food production. In Africa "at least 50 companies are involved in projects that have already put 3.2 million hectares of land into bio-fuel production cutting food supplies, and increasing dispossession and landlessness."[171] Adding to soil degradation, desertification and erosion, consider a 1-meter rise in ocean levels which, according to the IPCC, will wipe out a third of the land currently used to grow crops.

In the next few decades the amount of food produced must increase by 70% given an expected population of over 9 billion[172]. This will have to occur with less land, less fertility in the soil, less water (in quantity and quality), less *predictable* water, declines in crop yield due to climate change, and declining harvest from the ocean. Our inability to respond to these challenges is manifested in the increasing cost of food: "The onset of the global slump briefly arrested the escalation of food prices that had produced a wave of riots in 2008. But now, as the crisis changes form, they are on the rise once again and reaching unprecedented heights. Indeed, by late 2010 the UN Food and Agriculture Organizations' food price index hit an all-time high, after rising a staggering 32 per cent in the last half of that year."[173] As the U.N. World Food Programme has said, "A hungry world is a dangerous world. Without food, people have only three options: They riot, they emigrate, or they die. None of these are acceptable options."

This is not to argue that the earth cannot sustain 9 billion people. The point is that 9 billion people need food, which needs fertile soil, which needs water to grow the food, which needs energy inputs to plant the seeds, to pump the water, to harvest the food, and to transport the food around the world to the global population. And the amount of arable land, water and energy available to do this will depend a lot on Lifestyle, Organization and Technology. We have argued that Organization is unlikely to respond in a timely way - why should it when there is good money to be made? We have argued that most people will not change their Lifestyle, like reducing meat consumption and urban sprawl. And we have argued that Technologies like industrial farming and GMOs have been exploited to their potential and may in fact become detrimental to productivity due to changes in the future compared to low-impact organic farming and the cultivation of species of grains that have evolved to the geography and climate over millennia. In other words, we are unlikely to respond to food scarcity, and we are unlikely to be able to do so, even if we wanted to, given the twin carbon challenges of peak oil and climate change.

With respect to climate change: "The data point strongly toward a worldwide decrease in crop productivity if global temperatures rise more than 5°F (2.7°C) - well within the range of current predictions - although crop yields from rain-fed agriculture could be reduced by half as soon as 2020. … In addition, the rise in temperature may already be adversely affecting some crops - with higher night temperatures increasing nighttime respiration by rice (and perhaps other crops), resulting in the loss of metabolic energy produced by photosynthesis during the previous day."[174] And the prospect of responding to food scarcity with the advent of peak oil is unlikely, as we are essentially eating oil.

"In October 2009, Luc Gnacadja, executive secretary of the United Nations Convention to Combat Desertification, reported that based on current trends close to 70 percent of the land surface of the earth could be drought-affected by 2025, compared to nearly 40 percent today."[175]

John Bellamy Foster

Eating Oil

I keep dying in a dying age.

Becket, How It Is

We have already pointed out the threats to food security due to degradation and erosion of the soil, the increasing reliance on herbicides, pesticides and fertilizers, the unsustainable demand for surface and groundwater for irrigation, the failure of GMOs to increase food supply in the long term, the risk of climate change on food production (and land loss due to rising oceans), and the deleterious manipulation of food prices by oligopolistic corporations in food production and distribution. Another threat that has not been discussed is the reliance of food production on fossil fuels and the consequent impact on climate change. It is said that agriculture directly contributes over 13% to greenhouse gas emissions[176], whereas if all agricultural activities are considered, modern industrial agriculture "is responsible for 25% of the world's carbon dioxide emissions, 60% of methane gas emissions and 80% of nitrous oxide, all powerful greenhouse gases."[177] What is discouraging about these facts is that any effort to address climate change will have to address agricultural practices including meat production and the use of fossil fuels - and remember just how much 'free labour' we currently get from our use of fossil fuels in agriculture.

Fossil fuels are used to cultivate the land; to plant the seeds; for manufacturing fertilizers, pesticides and herbicides and the energy required to spray these chemicals; to harvest the food; process the food; refrigerate the food; and transport the food.

For example, to produce fertilizers alone, "One ton of ammonia requires about 33,000 cubic feet of natural gas to supply the hydrogen required."[178] A recent study of Canadian farms gave the following breakdown of energy use: "31 percent for the manufacture of inorganic fertilizer; 19 percent for the operation of field machinery; 16 percent for transportation; 13 percent for irrigation; 8 percent for raising livestock; 5 percent for crop drying; 5 percent for pesticide production; 3 percent miscellaneous."[179]

It is said that we "use 10 calories of fuel energy to produce one calorie of food energy."[180]

This reliance on fossil fuels in industrial agricultural practices not only adds to greenhouse gas emissions, it contributes to soil degradation which, in turn, results in greater reliance on fossil fuel: "due to soil degradation, the increased demands of pest management, and increasing energy costs for irrigation ... modern agriculture must continue increasing its energy expenditures simply to maintain current crop yields. The Green Revolution is becoming bankrupt."[181] Similarly, it is argued that soil erosion increases the need to replace essential minerals in the soil: "Average soil loss per hectare in the [United States] is now reported to be 13 metric tons per hectare per year (t/ha/yr). This means that an estimated 55 kilograms (kg) of nitrogen and 110 kg of phosphorus and of potassium (all essential nutrients for plants) are lost per hectare per year. To replace these nutrients requires about 880,000 kilocalories (kcal) per hectare for nitrogen and 440,000 kcal each for phosphorus and potassium per hectare per year. The annual total energy, just for these lost fertilizer nutrients, is 1.6 million kcal per hectare. This is about 20 percent of the total energy input that goes into producing a hectare of corn grain."[182]

So, just how much energy is used to grow our food? Producing "one tonne of maize in the US requires 160 litres of oil, compared with just 4.8 litres in Mexico where farmers rely on more traditional methods. In 2005, expenditure on energy accounted for as much as 16% of total US agricultural production costs, one-third for fuel, including electricity, and two-thirds indirectly for the production of fertilizer and chemicals."[183] As noted, energy replaces labour, and raising corn and most other crops by hand "requires about 1,200 hours of labor per hectare (nearly 500 hours per acre). Modern mechanization allows farmers to raise a hectare of corn with a time input of only eleven hours, or 110 times less labor time than that required for hand-produced crops. Mechanization requires significant energy for both the production and repair of machinery (about 333,000 kcal/ha). About one-third of the energy required to produce a hectare of crops is invested in machine operation."[184]

So, unless we are planning to repopulate the farms, we will continue to rely on diminishing amounts of energy to feed the global population - energy that continues to contribute to climate change when used. Ultimately, each person in the United States, on a per capital consumption basis, "requires approximately 2,000 liters per year in oil equivalents to supply his/her total food, which accounts for about 19 percent of the total national energy use. Farming - that portion of the agricultural/food system in which food is produced - requires about 7 percent and food processing and packaging consume an additional 7 percent, while transportation an preparation use 5 percent of total energy in the United States."[185]

This dependency on energy in food production, and for the manufacture of petroleum-based fertilizers and pesticides, means that peak oil is "peak soil."[186] In other words, using energy to make up for declining soil fertility, for declines in arable land, for declines in accessible fresh water for irrigation, and for the impacts of climate change can be sustained only as long as there is energy to do so. Peak oil is peak food. Peak food is peak population (even if we were to respond to reducing our taste for meat and our conversion of food into fuel - though the latter is self-limiting, as we will no longer be idling in the drive-through of a fast food outlet).

Burning Fossils Fuels to the 'Last Hundredweight'

Max Weber, not typically considered a critic of capitalism, observed

> that colossal universe that is the modern economic order, founded upon the technical and economic bases of a machinist-mechanical production which, through its oppressive constraints determines now, and will continue to determine, the lifestyle of all individuals - and not just economically active individuals - precipitated since birth into the cogs of this machine, until the last hundredweight of fossil fuel has been used up.[187]

Let's pretend that we actually chose to replace our fossil-fuel sources of energy with renewable energy. To do this we would need to divert fossil energy currently used for other 'needs' to the production of renewable energy technologies. Assuming that fossil energy is being produced at peak rates, energy consumption becomes a zero-sum prospect - one side gaining necessarily means the other is losing. Who is going to go without? The situation gets worse as we invest more energy just to produce more energy - as the rates of energy production decline both naturally (the bell curve) and due to the greater reliance on low-EROEI energy generation. More and more energy users have to go without simply in order to invest up-front in renewable energy which will produce energy slowly over a long period of time. By way of analogy, if energy were money, it would be difficult to get someone to invest in an industry with a low rate of return, like renewable energy, when there are still opportunities to profit from high return prospects like fossil fuel energy.

Wouldn't it have been nice to have invested in renewable energy technologies when we had a surplus of fossil fuel production? (Sniffle, gulp, gulp ...). To make the transition to renewable energy technologies when fossil fuel production rates peak means that we would have to greatly redirect energy from current 'needs' to future prospects for energy generation. This is just like the scenario in which we might have to go hungry during the winter to save our seed potatoes for planting next spring. The goal is to produce enough renewable energy to continually replace the generating capacity of the technology (the energy required to replace itself), and to provide for our energy needs while reducing life-cycle emissions to mitigate the worst impacts of climate change.

To divert this initial energy to renewable technologies production would, therefore, require that we reduce our consumption of energy that sustains our expected quality of life, or that we achieve much greater efficiencies in energy use. Since we have agreed that peak production rates of fossil fuel energy will limit any probability of increasing production, the opportunity to increase fossil fuel production in the short term is denied to us. There is a potential to realize greater efficiencies, but most would require a significant capital (and energy investment) upfront. Significant gains could be realized in some sectors, but in a competitive market that presumes the efficient use of resources, huge efficiency improvements across all sectors is quixotic. There is also the Rebound Effect (or The Jevons Paradox) that has to be considered, in which efficiency gains result in even greater consumption. This leaves the option of a reduction of consumption so as to divert energy to the production of the renewable technologies that would be required to replace the declining availability of fossil fuels. This begs the question: What would it take for people to reduce the consumption of energy, which is directly correlated to their consumption of goods (and standard of living)?

This reduction in consumption could be voluntary, where the majority of world-citizens realize the implications of peak production of fossil fuels, and the threat of global warming and the associated environmental impacts like biodiversity and food security, and social impacts like mass human migration. The citizens of the world would voluntarily restrict their consumption of goods, live in smaller homes, eat locally, stop travelling, and distribute their wealth to assist others more affected by the impacts of past behavior. Laughable? Indeed. Let's look at the current state of the genus *homo* and species *sapiens sapiens*.

... and we wouldn't do anything about it if we could.

Waste a Lot, Want a Lot

A pox on bad. Mere bad. Way for worse. Pending worse still.
First worse.
Mere worse. Pending worse still.

Beckett, Worstword Ho

As money is invested into new capital, more people are hired as wage earners, and this generates more surplus that must be reinvested (or consumed). The surplus reinvested into capital creates more commodities that must be purchased by wage earners. As capital replaces labour, the rate of production tends to outstrip the ability of wage earners to consume (overproduction). A reduction in demand (underconsumption) reduces the prices of the commodity and, therefore, reduces surplus (profit). If profits diminish, investors will look for other sectors that provide a greater rate of return on their investment.

> Confronted with a downward shift in final demand, monopolistic or oligopolistic firms would not lower prices (as in the perfectly competitive system assumed in most economic analysis) but would instead rely almost exclusively on cutbacks in output, capacity utilization and new investment. In this way they would maintain, to whatever extent possible, existing prices and prevailing profit margins. The giant firm under monopoly capitalism was thus prone to wider profit margins ... and larger amounts of excess capacity than was the case for a freely competitive system, thereby

> generating a strong tendency toward economic stagnation.[188]

In other words, to avoid stagnation, surplus must be effectively absorbed by the economy. In general, surplus can be absorbed in the following ways: it can be consumed, it can be invested, and it can be wasted.[189]

The first mode of absorption (consumption) is limited, as the people with wealth are motivated by increasing their savings through reinvestment and, furthermore, there seems to be a limit in the ability for the rich to consume luxury goods. It has been suggested that when the income share of the top 1 percent of wage earners reaches 25% of the total in the United States, recession follows - money is simply not being recirculated effectively enough, as the rich run out of those things to buy that will stimulate economic growth.[190]

The second mode of absorption (investment), results in even more production and, therefore, more surplus to absorb, making the problem worse. As a result, the third mode (waste) has become the contemporary mechanism of surplus absorption. The first mechanism of 'waste' is unutilized capacity. The general rule is: the surplus that is not absorbed is not produced - in other words, in times of stagnation, prices do not go down (there is very little downward elasticity in commodity prices), there is simply a reduction in production or, conversely, an overcapacity in production. For example, in 2002 manufacturing capacity utilization was at 74% in the United States, meaning that 26% of the available capacity was not being used in production.[191] This is 'waste' by not producing the surplus in the first place. When demand can be easily met by this unused capacity, there is no incentive to invest, and with a heavily indebted population, economic uncertainty, and high levels of unemployment, greater demand is not expected.

As discussed, underconsumption and the lack of investment opportunities for accumulated profits are two important factors in the tendency for an economy to stagnate. Underconsumption describes the limit of wage-earners (and those with savings) to buy commodities. To find an outlet for excess savings (due to the lack of investment opportunities), money is loaned to individuals without savings (at interest rates that exceeds the opportunities for profit in the market with similar levels of risk) who may then purchase commodities. Wage-earners, in essence, use future earnings to purchase commodities today. Easy credit encourages this behaviour.

> In the US between 1950 and 2005, for example, the total outstanding debt (household, business, government, and foreign) has increased 86 times compared with 40 times increase in GDP. ... According to the Federal Reserve statistics, as of the fourth quarter of 2006 the total outstanding debt for the entire US economy stood at $44.55 trillion: $14.13 trillion from the domestic financial sector, $12.82 trillion from households ($9.67 trillion in mortgages and $2.43 in consumer credit), $9.99 trillion from businesses, $4.88 trillion in federal government debt, $2.00 trillion in state and local government debt, and $1.72 in foreign-owned debt.[192]

The growth of consumer debt (in addition to public and corporate debt) is indicative of the efforts to forestall underconsumption. "Bolstering the economic miracle of the last 50 years now appears to require continued access to cheap credit, low taxation levels, and the whipping up of material desires in the increasingly affluent middle classes."[193] There are, however, limits to the amount of debt individuals can sustain - so there are limits to this strategy to avoid economic stagnation.

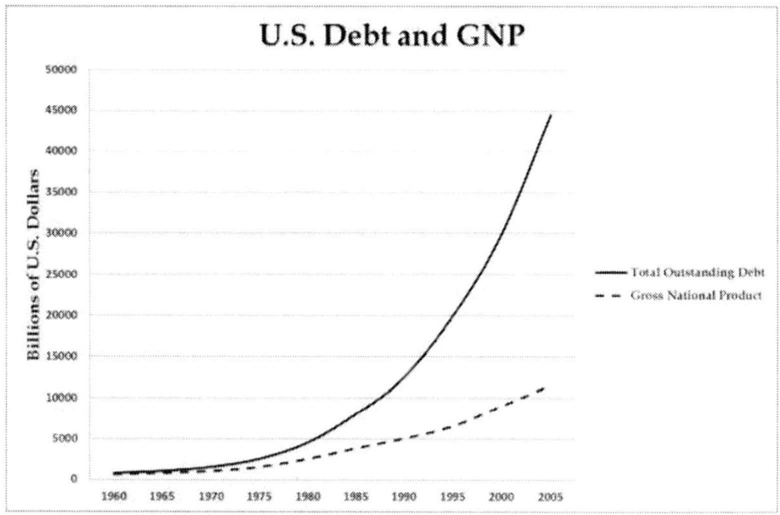

Figure 17 - U.S. Debt and GNP Growth

The United States national debt not including business debt is presently almost 16 trillion dollars (approximately $50,000 per person);[194] the average credit card debt is presently $15,000 per household, with an average interest rate of 18.9%;[195] in 2001, the ratio of household debt to disposable income exceeded one-to-one;[196] about 14% of disposable income is used to service debt;[197] the debt to profit ratio for the corporate business sector is approximately 10 to 1 (50% of the money borrowed by corporations in the late 1990s was to repurchase company stock so as to temporarily boost stock values).[198] The deficit for trade in goods and services in the United States exceeds 500 billion dollars.[199] "Between 1995 and 2002, the U.S. economy accounted for 96% of the cumulative growth in world GDP. The U.S. expansion has been financed by reducing domestic savings, raising the private sector debts to historically unprecedented levels, and running large and ever-rising current account deficits. The process is unsustainable."[200] What is worse is that we are borrowing all this money to create crap that we are convinced to consume and quickly throw away.

> Note how many of the leading modern economic historians equate consumerism not with wealth creation and societal growth, but with inflation and the decline of citizenship. Why? Because there is a constant surplus of goods that relate neither to structural investment nor to a concept of economic value, let alone to societal value.[201]

Once the opportunities to expand production (by boosting consumption) have been exhausted, excess surplus may be used to speculate in the stock market, or used by firms to integrate (horizontally or vertically) with other firms. Speculation as a mechanism of 'waste' inflates economic measures of performance "but adds nothing real to the economic wealth of the society."[202] Acquiring other companies provides an opportunity to invest surplus value, and these mergers (as they effectively converge into monopoly) tend to eliminate the supply side of the supply and demand relationship - and without competition, prices (and profits) are controlled by the producer. Additional surplus generated by the larger corporation, however, exacerbates the problem of finding investment opportunities. Furthermore, mergers allow large (and growing) firms to control the rate of innovation and, thus, reduce the rate of obsolescence of capital (due largely to the elimination of competition). Reducing the rate of capital replacement further increases profits and, again, exacerbates the problem.

This trend may be illustrated by the financialization of the economy which relates to the increased emphasis of finance, insurance, and real estate (FIRE) in the economy, which have grown relative to value-adding activities in manufacturing. The financialization of the economy adds no real value – it is a form of speculation in the desperate effort to find places to invest surplus. The recent housing bubble in the United States is clear testimony of the speculative character of FIRE and the blindness (ideology) of the economy as it relates to value-added activities. Speculation followed by massive devaluations is an effective form of waste that can mitigate over-accumulation.

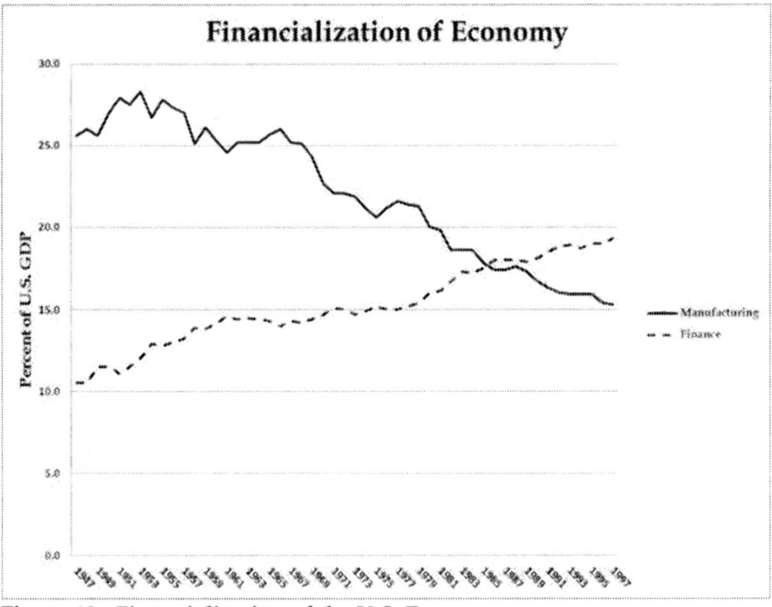

Figure 18 - Financialization of the U.S. Economy

The chilling of technical innovation has the further effect of reducing the likelihood of a world-changing contribution to our economy. Common examples of innovation that have stimulated economic growth for long periods of time are the railroad, the automobile, and the computer. These types of innovation transformed how the world worked and revitalized the economy as it made old technologies and production methods obsolete (Schumpeter's 'creative destruction'), and provided a new source of investment opportunity (which absorbed the excess surplus for a while). Are there future innovations with the potential of absorbing surplus? Some suggest that nanotechnology has a potential world-changing impact, but monopoly control of intellectual property threatens widespread application and, therefore, the sort of growth that can effectively absorb surplus.

Another approach to reducing the effect of underconsumption is to find new consumers who are not currently involved in the cycle of production as wage earners. As we have argued, wage earners in the production cycle produce surplus, and it is precisely the surplus of surplus that causes stagnation of the economy. Finding (or creating) non-producing consumers is paramount. One approach is to sell commodities to economies that do not have the profit motives that create surplus in the first place: the globalization of capital is a manifestation of this tendency,[203] as regions that have existed with non-capitalist or self-sufficient economies in the past are absorbed into the global capitalist economy. These sorts of cultures are fast disappearing. Another approach is to 'waste' surplus by investing in non-producing industries.[204] Common 'wastes' include advertising, the military, education, health and welfare, and government. The term 'waste' should not be interpreted as being useless - it simply means that it does not directly contribute to production and the direct generation of more surplus. This sort of 'waste' is effectively a safety relief valve for excess surplus in the economy.

In general, however, there must be a reason for those with surplus money to 'invest in waste' - the reasons for this type of investment typically support the processes of production. For example, the advertising industry absorbs a considerable percentage of the excess surplus.

> In the efforts of monopolists to enlarge their sales without jeopardizing the existence of extra profits we find the fundamental explanation of the enormous development of the arts of salesmanship and advertising which is such a striking characteristic of monopoly capitalism. ... packaging and labeling, ... staffs of salesmen and publicists, ... newspapers, magazines and radio.[205]

Advertising creates a demand for the commodities produced (some of them with little or no value for society), and at the same time creates more consumers (the people making a wage working in this industry) who don't directly add to the surplus value.

> Thus part of the goods which are annually produced and which are called wealth, is, strictly speaking, waste, because it consists of articles which, though reckoned as part of the income of the nation, either should not have been produced until other articles had already been produced in sufficient abundance, or should not have been produced at all. And some part of the population is employed in making goods which no man can make with happiness, or indeed without loss of self-respect,

> because he knows that they had much better not be made, and that his life is wasted in making them.[206]

Health and welfare services also contribute to the absorption of surplus and, as a benefit, create a healthy and able workforce. The same could be said for educational institutions. The government plays a role in the taxation and distribution of surplus, while absorbing excess surplus in its own operations. The benefit of government to industry is the enforcement of a discipline that benefits the growth of the economy as a whole. Surplus is also used by the government for welfare, police and prisons, which are ostensibly used to control the population that has been excluded from either production or 'waste' industries. Finally, the military (as a form of waste) absorbs surplus in a most emblematic way.[207] The commodities of production are consumed in training and warfare, while creating another group of consumers that don't contribute directly to making surplus. One possible benefit of the military to industry is to secure and stabilize markets in other regions of the world that will purchase commodities.

Surplus is also used in apparently more philanthropic ways, such as promoting the arts, sports and entertainment (circus); promoting academic research (creating knowledge and innovation that can then be privatized to make more surplus); and supporting health and development in disadvantaged regions of the world (who we need as consumers of commodities).

The cycle is clear: Corporations make surplus by keeping wages low and by selling commodities dearly. The surplus is invested to hire more people to make more things we don't need. They keep wages low by intensifying work, by capitalizing and putting people out of work, and by keeping the cost of necessities low. But low wages curb consumption, so easy credit is created so people spend money they haven't earned yet, making them more dependent on their jobs creating an opportunity to keep wages low, particularly as more people are 'set free' from the workforce by capitalization. When the limits of debt and consumption are reached, surplus moves to underdeveloped regions and is invested to make things for the people who have been drawn into wage labour. When the surplus has been used to envelop the globe in one economy, and when people have borrowed their future earnings, the gig is up. Before this, you may be relieved to hear, we will have destroyed our ecology and consumed our minerals and resources until the 'last hundredweight of fossil fuel has been used up'. And even before this, the people who have been discarded by the process, particularly those people who once derived some benefit from the process, will likely tear the system to the ground.

One has to ask: What sort of people would perpetuate such a system? What lies must be told to convince us to accept it?

Not Believing in Climate Change – A Rant

They fill the children full of hate to fight an old man's war and die upon the road for peace.

Tom Waits, Road to Peace

To illustrate what type of people we are, let's examine at the efforts of the climate contrarians. Why do so many people deny global warming and the impending challenges it poses to our future? We have witnessed three successive stages of denial: "(1) the denial altogether of the planetary ecological crisis (or its human cause); (2) the denial that the ecological crisis is fundamentally due to the system of production in which we live, namely capitalism; and (3) the denial that capitalism is constitutionally incapable of overcoming the global ecological threat - with capital now being presented instead as the savior of the environment."[208]

For the past couple of decades, since the first IPCC report, the denial industry in North America has been steadily and resolutely trying to sow confusion. The motives of the (predominately energy[209]) industries are clear - their existence is premised on burning lots of fossil fuels. They say that they don't believe in climate change - as if science is a belief system. This is like saying you don't believe in gravity, or that you don't believe in subatomic particles. Either the science supports the evidence (as in the case of climate change) or it does not. The second most commonly touted position is that it is 'arrogant' to think that mankind could possibly affect the climate. Wow. Is it arrogant to think that we are burning huge holes in the ozone that protects us from damaging radiation from the sun? Is it arrogant to think that we have fished the apparently limitless oceans to near extinction? Is it arrogant to think that we could essentially wipe out higher life on earth with nuclear war? It is clear that by emitting millions and millions of tons of known greenhouse gases into the atmosphere, at rates well above the ability of the earth to naturally absorb, humans are quite capable of disrupting natural climate systems. Hell, if phytoplankton can do it, humans, with a little effort, sure can: "There is plenty of evidence that even some individual species have a measurable global impact. In the most notable example, the oceanic phytoplankton, composed of microscopic, photosynthesizing bacteria, archaeans, and algae, is a major player in the control of world climate." Indeed, if there is any arrogance being displayed, it is the conceit of people who would dismiss scientific consensus in any field of knowledge.

Or, to look at arrogance from another perspective, consider the hubris of a species that thinks it can reproduce what the complexity of nature already provides – to become, what James Lovelock has called, the 'planetary maintenance engineer'. This from a species that has shown little ability to maintain itself in any other way for any significant period of time: "The most extreme expression of human arrogance ... is the idea that human beings can save the planet from environmental destruction. Because they are killer apes, that is, by virtue of a naturalized version of original sin that tends them towards wickedness and violence, human beings cannot redeem their environment. ... *Homo rapiens* is ravaging the planet like a filthy pest that has invested a dilapidated but once beautiful mansion."[210] To believe that killer apes can save themselves – now, that's arrogant. (Perhaps we went a little too far on that one.)

Another common perspective from the deniers is that the science is uncertain, that the models aren't perfect. Duh. We're talking about modeling a vast planet with hundreds of millions of interacting systems - to think we could do that is truly arrogant. But uncertainty in science and gaps in the model do not mean that the tendency cannot be predicted. Scientists don't understand quantum physics, but this does not make them doubt the model's predictive value. We don't understand what causes gravity, but we can still put a satellite into orbit around the earth.

Take a football game for example. You have been asked to model a professional team playing a high school team in Fort Bridger, Wyoming on November 6th. You know the rules of the game and some constraints like how far a ball can be thrown or kicked, and how fast a human can run. (Just like we know that greenhouse gases absorb heat energy, the bands of radiation within which these gases interact, the mechanisms of absorption by other systems, and so on). You can measure some of what you consider to be the main variables for team performance - these are the statistics collected by any avid sports follower: Average passing and running, the size of the players, experience, and performance in different locations or weather conditions. (Just like we can measure average global temperatures and carbon dioxide concentrations, ocean acidity, species extinction, late springs, reductions in precipitation, and so on). Now by putting the rules of the game together with the most significant statistics and observations, you can model the game. Four quarters, average time per play, average distance the ball moves per play ... voila, you have a model. It might not play out exactly as you expected, after all, there is a lot of room for uncertainty, but the outcome of the game will be quite predictable. If you don't believe that, just ask your bookie if he makes any money on sports gambling.

Now, along come the people who really want the high school team to win - their beautiful child is the captain, and even the thought that those nasty modellers have dismissed the chances of her team winning against the professional team are unacceptable. "Did you consider that the high school team has never lost a game in Fort Bridger, Wyoming?! Did you consider that the average foot size of the high school team is the same as that of the professional team?! I read that one member on the professional team is on a diet, so your average team weight is wrong! See, your model is deficient. I don't believe it."

And then there is the Crichton-argument: All of the scientists on earth working on aspects that relate to climate change are conspiring in a collective effort to gobble up those yummy grants that hang from trees everywhere. Really? Find me a scientist that has grown rich off of their climate-scientific endeavors … in reality, the more prominent ones have spent personal fortunes defending themselves from the venom of the denial industry. Let me ask: if you were a climate scientist wishing to access grants or to get rich on the speaker-circuit, would you choose the side opposed to the people that have all of the money? If money were a motive, I would look into the denialist camp and follow the threads to their funding sources. But what better defense than a good offence … a sophisticated offence like, "I know you are but what am I?"

The reason that the energy industry would wish to obfuscate the scientific consensus on anthropogenic climate change is quite evident … I just had to get the obvious objections off of my chest. The real question is: why do so many people allow themselves to be influenced by these arguments? Fundamentally, it is because climate change is an ethical issue. There are a number of reasons for our unwillingness to rationally react to climate change: 1) the causes are separated from the effects; 2) there are a great number of sources of greenhouse gas emissions; and 3) we have extraordinary inertia within our guiding institutions (Organization).

The separation of cause and effect is manifested spatially and temporally. Spatially, the impact of an emission in Calgary impacts the lives of the people of Katmandu. And the current impacts of climate change are significantly 'backloaded' temporally from the industrial growth of previous generations of emitters. It has been argued that prosperous nations have already used up their emission-quota over the past two centuries while creating their current wealth, and that the present generation must stop emitting to allow for other nations to use their quota. (You can see why that argument would be an easy sell in developed nations). The spatial and temporal separation of cause and effect of global warming make a fair resolution untenable.

In addition to spatial and temporal separation, there is a class separation in which the wealthy are disproportionally responsible for emissions. Yes, the same wealthy that fund the denial industry, manage profit-seeking corporations, and sit in the seats of government.

> According to a study by Princeton ecologist Stephen Pacala, the world's richest 500 million (roughly 7 percent of the world's population) are currently responsible for 50 percent of the world's carbon dioxide emissions, while the poorest 3 billion are responsible for just 6 percent.[211]

The second problem is that the combustion of fossil fuels occurs everywhere, from large to small, from belching coal-fired electricity generating facilities to the personal scooter. This relates back to the impact of lifestyle on the environment. Even if there were institutional efforts to reduce emissions, how could so many sources be effectively monitored and controlled? But, not to worry, there is no risk of this because there are no world-wide institutions that are capable of managing an effective effort to eliminate emissions. Even the widely discussed polluter pays principle (PPP) for addressing climate change would be ineffective because of the responsibility of earlier generations of emitters who can no longer pay and were ignorant of the future implications, and the unwillingness (not to mention institutionally unenforceability) of current citizens of polluter-countries to pay for these previous generations, even though their current wealth is premised on these past emissions.

> It may be appropriate to sum up. I have argued that the PPP [polluter pays principle] approach to climate change is inadequate for a number of reasons. It cannot cope with three kinds of GHG, namely GHGs that were caused by: (i) earlier generations (cannot pay); (ii) those who are excusably ignorant (should not be expected to pay); and (iii) those who do not comply with their duty not to emit excessive amounts of GHGs (will not pay).[212]

And with respect to inertia, we have a 'democratic' political machine premised on gaining power and maintaining power. These efforts require a lot of money (and remember who has all of the money). But it also requires a lot of votes, so a concerted effort is required to manufacture consent. As such, the plutocratic machine requires enough of the electorate to align their values with industry to maintain power. Voila, the industry of denial is born. And, hence, the inertia of political efforts to react to the emerging challenges becomes obvious. Like we said earlier, challenges like energy security and the impacts of climate change require a significant commitment to planning well in advance of the imminent crises. Planning in the absence of crises is anathema for politics and results in NIMTO ("Not-In-My-Term-of-Office").

Furthermore, as David Keith argues: "The sad fact is that the optimal strategy for each country is to get other countries to cut their emissions while each country does nothing. The sad fact is that if you spend a lot of money to cut emissions in your country, you're distributing the benefits of that cutting all over the world, but all the costs of the cutting are in your country. This is the way an economist would think about it. Another sad fact is that one of the things that people talk about a lot now, more and more publicly, is the difference between mitigation - cutting emissions - and adaptation: dealing with the climate change. But when you think about this from the perspective of a national government, if you spend money on adaptation in your country, you know the money will be spent in your country, and the benefits will be there. If you spend money on mitigation, those benefits are being spread around the world."[213] In other words, there are serious barriers to 'selling' mitigation efforts, when adaptation to already-existing manifestations of climate change will appear more directly as benefitting the taxpayers.

Imagine the Titanic racing through the ice flow. The Captain thinks, 'the ship is unsinkable, and it has to arrive on time on this maiden voyage.' A few of the crew stand beside the life-rafts hand wringing - 'wouldn't it be prudent to slow down a bit?' they think. The ship engineer calculates the likely speed of impact and the strength of the cold steel - 'unsinkability is only an ideology,' he thinks. The passengers see the hand-wringing crew standing along the rail, and then gaze upon the captain who is confidently standing on the bridge up high. The engineer is below decks making more calculations that continue to support his concerns. At first a large number of the passengers wait on deck. They generally think it would be a good idea to slow down, but they admit that they don't know a lot about ships. The expert engineer has not emerged from below decks to say anything that contradicts the appearances. Nothing changes for *one whole hour*. The music and scent of food drifts through the chill air. The passengers begin to migrate back to their friends and their drinks in the warmth of the glamorous ballroom. Seeing others enjoy themselves makes them forget. Thud. "Oh, me. Oh, my. We should have slowed down! We should have … glug, glug, glug."

What is the definition of moral corruption?

Moral corruption: distraction, complacency, unreasonable doubt, selective attention, delusion, pandering, false witness, hypocrisy.

Hmmm.

Moral Corruption of the Corporation

I've got no strings
To hold me down
To make me fret, or make me frown
I had strings
But now I'm free
There are no strings on me

Ned Washington, I've Got No Strings

When we invited corporations into the human family in 1819 (first through the Fourteenth Amendment, then the Fifth Amendment, and in 2010 in a First Amendment decision in the United States) we invited a very rich and immortal group of 'people' to participate in our political process. 'Hi-ho the-me-ri-o, that's the only way to go' - when we cut the strings, the little boy could do anything! Oh, but never, never, never, never, never, never, never, never, never, should he ever tell a lie …

Can a corporation be moral? Can it be morally corrupt?

Consider the energy industry – that is oil, gas and coal. The latest global evaluations on subsidies to the fossil fuel sector suggest values exceeding $300 billion each year – "a total that probably under-counts the full expense since it includes primarily easy-to-measure direct transfers, but captures more complex subsidy mechanisms in tax, credit, insurance and regulatory interventions much more inconsistently."[214] These are 'no-win' subsidies that divert money from other priorities, particularly social priorities, and they lead to excessive consumption of energy which amplifies pollution and energy insecurity worldwide. Once a subsidy has been given to these special-interest corporations, investment congeals around the subsidies making them difficult to change, even though they have lost whatever justifying purpose they may have once had.

What is more disturbing is that we subsidize an industry in which the producers seem to be doing okay on their own. The Big 5 oil companies (BP, Chevron, ConocoPhillips, ExxonMobil, and Royal Dutch Shell) made a record high of $137 billion in 2011, up 75% from 2010, and they have accumulated over one trillion dollars in profits over the preceding decade.[215] These are big numbers, though the apologists for subsidizing very profitable corporations say that the 'profit margin' – that is, the net profits divided by sales – is relatively low compared to other industries at 7.9%. They argue that the money is being made on shear volume, rather than price-gouging. Regardless, the question to be asked is if subsidizing these corporations is really necessary?

But this isn't really the morally corrupt part of the story, nor is it the reason why 'we wouldn't do anything even if we could'. These corporations who profit so handsomely from nature's beneficence, and who gracefully accept the general munificence from the generous taxpayer, are the same corporations that bully rivals in the sphere of making energy. The Guardian reported[216] that a number of conservative think tanks and lobbyists, supported by major corporations in the fossil fuel industry, are using 'subversion of message' to build a national movement of wind farm protesters. That's right – using subsidized profits to make more difficult the expansion of our most cost-effective alternative to fossil fuels. The report says: "It suggests setting up 'dummy businesses' to buy anti-wind billboards, and creating a 'counter-intelligence branch' to track the wind energy industry. It also calls for spending $750,000 to create an organisation with paid staff and tax-exempt status dedicated to building public opposition to state and federal government policies encouraging the wind energy industry." Think tanks and lobby groups connected to oil and coal interests are organizing strategic sessions to advance a concerted attack on the clean energy industry.

The main arguments used by these well-funded bullies against clean energy, and wind power in particular, is that it is a global conspiracy to make people 'pay more and use less', and it is a blight on Nature (as if, by comparison, oil pipelines and roads sectioning up the natural landscape have superior aesthetic value). In 2008 a wind farm near a remote Scottish island was blocked because the view would be destroyed. As Timothy Morton has said, this is "truly a case of the aesthetics of nature impeding ecology." In other words, we would rather see Nature (as it dies) than preserve the ecology that sustains it. As an alternative viewpoint, Morton waxes poetic to describe wind turbines "as environmental art. Wind chimes play in the wind; some environmental sculptures sway and rock in the breeze. ... an ethical sublime that says, 'We humans choose not to use carbon'."[217]

Short of a collective awakening to the vision of wind turbines, solar panels, solar concentrating power stations as art and the ethical sublime, it is unlikely that corporate resistance to these cleaner technologies will abate.

"I see in the near future a crisis approach that unnerves me and causes me to tremble for the safety of my country. Corporations have been enthroned, an era of corruption will follow, and the money power of the country will endeavor to prolong its reign by working upon the prejudices of the people, until the wealth is aggregated in a few hands, and the republic destroyed ..."

Abraham Lincoln

Media

> The best lack all conviction, while the worst
> Are full of passionate intensity.
>
> Yeats, Second Coming.

What about the corporate media? Surely, in a free and democratic society, these champions of balance and evidence and boldness would inform the public of the dangers of global warming and other emerging challenges to our way of life?

> But at the end of the day a huge amount of blame has to be laid at the feet of what in the Internet age is known as "mainstream media." Whether through inadvertence, understaffing, or, in certain outlets, an actual intent to misinform, many major media outlets in North America and, to a lesser extent, in Europe have failed to inform their readers and listeners about what is surely the most important and dangerous environmental issue in the history of humankind.[218]

Could a group of people, who we have entrusted with the task of informing the public with the (small 't') truth, have failed in their mission? Was it understaffing; was it public disinterest in a 'disposable beat'; or was it intentional?

Part of the problem is the monopolization of media by large corporations which reduces competition and, therefore, results in poorer journalism caused by the reduction of resources over time. Eric Pooley of Harvard provides an indispensable critique of the media in climate change reporting. He supports the observed assault of economic forces on journalism: "Some of the decline may be due to climate fatigue among editors, but some surely comes from industry-wide layoffs that degrade the ability of news outlets to cover any specialized topic. Major chains such as Cox Newspapers and Advance Publications are shutting down their Washington bureaus, and CNN recently dismissed its entire seven-person science-and-environment unit."[219] With reduced resources and fewer journalists, media relies more heavily on official sources and becomes more susceptible to well-funded output from public relations firms and ideological think-tanks.

Take, for example, the concerted efforts of a well-funded denial industry to create uncertainty around the science of climate change: "For a decade or more, the U.S. debate about anthropogenic global warming was prolonged by a well-orchestrated (and by now well-documented) disinformation campaign designed to sow doubt and delay action." The denial industry has effectively wielded the tool of fear-mongering around economic effects: "The terms of that debate have been defined by opponents of climate action who argue that reducing emissions would 'cost too much'." Pooley further suggests that the battle has been "fought over the short-term price of climate action and its impact on GDP, while overlooking an extremely important variable, the long-term costs of inaction and business as usual."[220]

As an example, the Keystone XL pipeline has been designed to deliver bitumen from the Alberta tar sands to a refinery in Texas, passing through many U.S. states, across important groundwater aquifers and sensitive natural areas. The rational argument against the Keystone XL has been environmental – pipelines leak and we can't afford to contaminate our diminishing sources of fresh water that support agriculture and, well, life. The media, however, decided to argue the economic concerns over jobs that may or may not be created during the project and afterward. Many credible analysts have suggested that the industry estimations of job creation are grossly over-estimated. But what coverage did the press give? A Media Matters[221] study shows that the broadcast media hosted or quoted people who supported the pipeline 79% of the time, with 7% of the time going to those opposing the pipeline. The ratio for print media was 45 percent and 31 percent, respectively. With respect to the topics covered, the broadcast media talked about jobs almost 70% of the time, and the environment 17% of the time. Again, the print media was more balanced with a 68% and 65% ratio, respectively. Even more tellingly, given the contentious data about job creation, the broadcast media never mentioned the criticism of industry-provided numbers, while the print media did so 25% of the time. The bias in reporting was clear - it was focused on jobs, when the argument was about the environment.

Another pipeline designed to deliver bitumen from the Alberta tar sands to China is being planned, and will span the province of British Columbia to a port near Kitimat. The pipeline crosses thousands of rivers, the lands of First Nations who are unanimously opposed to the project, and will fill supertankers. These supertankers must travel treacherous waters though seas that are considered necessary habitat for endangered humpback whales and other rare species. The coastal waters are essential for the livelihoods of First Peoples and the tourist economy of British Columbia (and Alaska). But we will risk it all for oil revenues. The Government of Canada under Stephen Harper has aligned with industry in a heavy-handed propaganda campaign implying foreign influence in environmental resistance and is currently rewriting legislation that environmentalists and First Peoples are using to hamper a quick approval process.[222] Where is the media?

A similar story can be told about climate change. Given the "vociferous arguments on both sides during the Bush years and disinformation coming from the highest levels of the U.S. government, many news organizations restricted their climate reporting to a facile balancing of opposing views on the subject, even though there was a scientific consensus that carbon emissions were warming the planet in potentially catastrophic ways."[223] McChesney discusses that "between 1995 and 2005 nary a single one of the nearly 1,000 referenced academic research journal articles on climate change disputed the notion that something fundamental, dangerous and influenced by humans was taking place. Yet our news media sources representing the interests of oil companies and other major polluters provided political significant official opposition to the notion that global warming was a problem or that pollution had anything to do with it. ... Over one-half of the 3,543 news articles in the popular press between 1991 and 2005 expressed doubt as to the existence and/or cause of global warming."[224]

Though Pooley outlines the economic effects of diminished journalism generally, and the confusion created by the denial industry, his argument is predicated on the position that the media *intends* to provide well supported evidence and balance in their reporting - you know, the type of balance that I have demonstrated in this book. He advises that the "press has an obligation to remain clear-eyed and skeptical because with the policy issues so complex and the stakes so high, we can't afford to get this wrong."[225] What Pooley fails to argue is that the media is owned and controlled by large corporations that have their own (obvious) interests in business as usual, and rely on advertising from other corporations to make their media investments profitable.

David Harvey might call this the influence of the Party of Wall Street: "The Party of Wall Street knows all too well that, when profound political and economic questions are transformed into cultural issues, they become unanswerable. … to create controversies out of all manner of issues that simply do not matter, and to propose solutions to questions that do not exist."[226] It is the creation of issues that don't matter or by transforming issues into unanswerable cultural problems that the corporate-owned media is able to deflect attention from really existing issues that do require solutions.

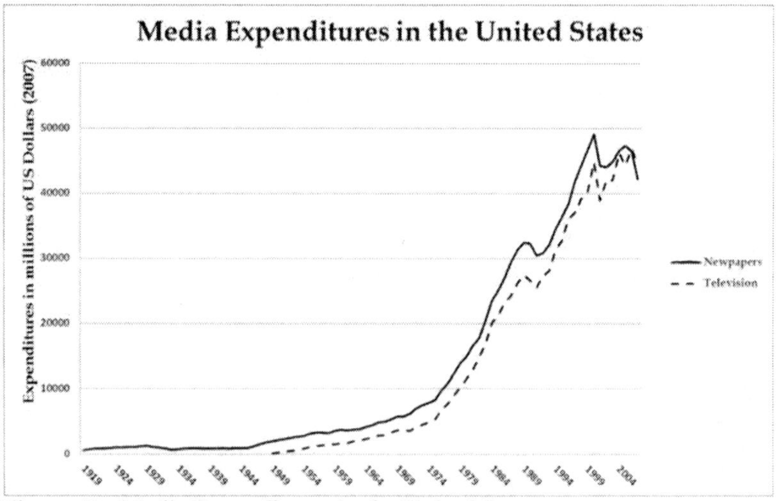

Figure 19 - Media Expenditures in the United States

In a 1992 survey conducted among members of the Society of American Business Editors and Writers, over 80 percent reported that advertising pressure was becoming a growing problem and 55 percent said that advertising pressure had compromised editorial integrity. A similar survey of newspaper city editors showed that 90 percent believed that advertisers attempted to influence the content of news stories, and more than 70 percent reported that advertisers had tried to kill stories at their paper[227].

It is difficult to dispute the influence of money in a for-profit industry: the chart on media expenditures shows the growth in 2007 dollars of advertising in newspapers and television over the last century. And, yes, you are reading the chart correctly if you see over $40 billion of ad money going to each of the two main media sectors. Worldwide, almost $700 billion is being spent in advertising - almost half being spent in the United States (another effective 'waste' industry to maintain the growth of capital). About $1000 is spent each year on each person in the United States[228] to convince them to buy crap they don't want or need. And that same $1000 effectively motivates corporate media to tow the line.

Education for Citizenship and Democracy …

You'ld better start swimming
or you'll sink like a stone.

Bob Dylan

Of course, if we had an education system that prepared future citizens to be able to intellectually and emotionally protect themselves from the onslaught of paid reportage, things would be better. But that is not the case. The influence of ideologues on state-mandated educational curricula is as pernicious as their influence in the media. Maybe worse, considering that expectations for balance and public trust is greater.

Since we already know that there is a scientific consensus on the mechanisms of global warming and the impacts of climate change currently being observed, one would think that educational curricula would be treating the topic as science – kind of like gravity, the cause of which we know even less about than climate change. Using a 'model bill' called the Environmental Literacy Improvement Act[229], states across the United States have mandated teaching 'balance' in the sciences, including climate change. A Lousianna bill legislates "an environment within public elementary and secondary schools that promotes critical thinking skills, logical analysis, and open and objective discussion of scientific theories being studied including … global warming"; a South Dakota bill calls for "balanced teaching of global warming in the public schools of South Dakota"; Kentucky … New Mexico … Tennesee … Okalahoma … Texas … all parrot the ALEC model bill to create controversy where there is none.

It should be said, loud and clear (even though I am writing), that we should be grateful that there are so many dedicated and effective teachers who can see through the bullshit. No matter how hard the state tries to crush their spirit, no matter how much funding the state diverts from education, no matter how zombified the students become, they keep trying to make the world better. I think a postage stamp should be dedicated to each and every one of them. Oh, yeah! (But, of course, the odds are against them.)

What is especially pernicious about the ALEC model bill is that it calls for 'balance', 'critical thinking', 'exploration of perspectives', 'open-mindedness', while asking 'not to include instruction on political action skills or encourage political action activities' with the sole intention of forcing teachers to try to present scientific controversy over climate change (and other scientific theories) where there is none. When the scientists are debating a controversy, sure; but when the scientists see no controversy, then there is none. Noam Chomsky has said for years that the best system of propaganda is one in which the population does all the work of suppressing undesirable viewpoints (what is left is the 'unknown knowns' called ideology). The media and education - through underfunding, through ignorance, through the influence of advertising income, or simply through complicity - are no longer providing the skills for citizenship or for understanding the environmental and social challenges that are emerging. This unpreparedness is reflected in the quality of governance.

"The advertising industry's primary task is to ensure that uniformed consumers make irrational choices, thus undermining market theories that are based on just the opposite. And public relations recognizes the benefits of undermining democracy the same way."[230] Noam Chomsky

Just How Corrupt Are Our Governments?

How many ways can you polish up a turd?

Tom Waits, Hell Broke Luce

This will be a short section, as there are very few people who dispute that governments serve capital. It has been written that (in 2011) there were over 12,000 paid lobbyists in Washington D.C., spending upwards of $3 billion to ply their craft[231] (this does not include direct campaign contributions, which have recently been deregulated for corporations). Top spenders include the US Chamber of Commerce, American Medical Association and Pharmaceutical Research & Manufacturers, and a list of major corporations including General Electric, Exxon, Boeing, AT&T, General Motors, and so on.

It should be noted that 'lobbying success' does not prove 'influence', as the government official may have already been inclined to work in the direction of the lobbying institution (... just kidding). An interesting study by Christine Mahoney of Syracuse University[232] shows that in both the United States and the European Union that the rate of success for lobbyists representing niche or sector industries was very high, in the range of 30 to 60%. More interestingly, when the data is disaggregated to the type of lobbyist, corporations in the United States had some success 89% of the time compared to 40% success rates for citizens or foundations. As for the type of success, 81% of the time the decision was to maintain the status quo.

A study by Yu & Yu shows that over half of former congressmen and senators become lobbyists for firms, having mastered the paths of influence.[233] Hill, Kelly, & Van Ness show that the market value of each lobbying dollar spent is roughly $200 – a very worthwhile investment. Maybe this is why the Big Oil companies need their subsidies, as they only receive about $30 for each dollar invested in lobbying.[234] These financial benefits may relate to reduced taxes; political connections and preferential treatment for government contracts; relaxed regulatory oversight (or greater oversight of competitors); relaxed financial regulations like investor-protection policies, market-entry policies, and financial investment policy; maintaining the status quo when challenged by changes to (environmental or social) standards; or maintaining open global markets through diplomacy or imperial force.

Guo cites an example of the 2007/2008 election cycle in the United States in which Citigroup Inc. spent $4.8 million on campaign contributions and another $7.6 million on lobbyists. After the financial meltdown, this same corporation received over $50 billion of tax-payer's money under the Trouble Assets Relief Payment program – a substantial return on investment. Under the same plan, each of us could invest a dollar and get $4000 back a year later. [235] Guo cites a study that suggests a 1% lobbying expenditure correlates to a lower tax rate between 0.5 and 1.6% - almost an instant payback, leaving the other benefits of lobbying and political contribution as free gifts. This is not to imply that corporations are 'buying political favour' (… still kidding).

The question remains: how are progressive changes to be made to avert or mitigate the many challenges facing civilization? It will not likely be through elected governments (or other forms of government), for reasons that are manifestly clear. What motive does the system of governance have to swim against the tide of corporate influence? This is particularly true in systems that don't discourage corruption and in systems where information is most readily accessible (and prettily packaged) from lobbyists and ideological think tanks. Easy information, easy money - don't expect change from above.

> And this is the second paradox: there is no democracy without true discourse, for without true discourse it would perish; but the death of true discourse, the possibility of its death or of its reduction to silence is inscribed in democracy. No true discourse without democracy, but true discourse introduces differences in democracy. No democracy without true discourse, but democracy threatens the very existence of true discourse.[236]

The maxim of Thucydides is that the strong do as they wish and the weak suffer as they must. This is not much different than economist Thomas Ferguson's 'investment theory of politics,' in which "we can understand elections to be occasions in which groups of investors coalesce to control the state, a very good predictor of policy over a long period."[237] We spoke earlier of the contradiction that business would promote anti-government sentiment while completely relying on government and the law to manufacture opportunities and to protect their interests. Chomsky offers in interesting reason why antigovernment campaigns have to be 'nuanced and sophisticated': this is because the 'architects of policy' understand "very well the need for a powerful state that intervenes massively in the economy and abroad to ensure they their own interests are 'most peculiarly attended to.' The goal of sophisticated business propaganda is to engender fear and hatred of government among the population, so that they are not seduced by subversive notions of democracy and social welfare, while maintaining support for the powerful nanny state for the rich – a difficult course, but one that has been maneuvered with considerable skill."[238] In other words, corporations 'invest' in governance that promotes its interest while maintaining a strong distrust of government within the general population lest they get the crazy idea that they can influence its direction through the democratic process.

Piven in *The New American Poor Law* summarizes this nicely: "Even putting aside the 'end-times' wonderings about climate change, food scarcity and nuclear disaster, the neoliberal juggernaut in the US is on a roll, calling for union busting state laws, ever greater clawbacks in the form of tax cuts for business and the affluent, and budget cuts for everyone else, but especially for the worst-off, coupled with court-ruling that open the electoral system even wider to the influence of money while other changes are introduced into electoral administration that make voting more difficult for the growing population of naturalized citizens, youth, and the poor." [239] He goes on to say: "The growing complexity of governance makes the public acutely susceptible to propaganda, especially propaganda that takes the form of stories that give simple explanations, usually point to particular villains who are ostensibly to blame for what is going wrong in their society."[240] Of course, the villains are the shiftless and promiscuous poor indulged by the largesse of a too-caring and too-generous society. You can also throw in villains like animal rights groups, environmentalists, and anti-capitalists (as the Canadian Government did in 2012, implying that these groups were threats to public security along with white supremacists and terrorist groups).

A corporate-supported structure of government not only suppresses education, it is complicit in maintaining the ideology of neoliberalism. The damage this does is incalculable: what is the social cost of stifling debate, of excoriating groups like animal rights, environmentalists, and anti-capitalists who try to maintain an ethical gaze on elected decision-makers? Al Gore (once complicit in the system) summarizes this: "Good ideas in the minds of men and women who cannot afford the price of admission to the public forum are then no longer available for consideration. When their opinions are blocked, the meritocracy of ideas that has always been the beating heart of democratic theory begins to suffer damage."[241]

How Many Voters …?

Have you heard the joke about voters? No? Well, then … How many voters does it take to change a light bulb? None – voters have never been able to change anything.

I think one of the greatest deceptions advanced by corrupted governments is that voting matters. In fact, it is presented to be the only thing that *does* matter in a democracy. Have you heard people say things like – if you don't vote, you can't complain about the government. They say this as if there actually was a choice given to change the government through voting (even though, no matter who you vote for, a government always gets in). You hear the wisdom proffered through the media that to vote is to exercise your democratic right and duty. When did voting become the essence of democracy? That makes about as much sense as saying the exclamation mark is the essence of the sentence!

!

There are a number of problems with the political process today: first, there is the protean definition of democracy. During election time, democracy is the competitive process by which the electorate chooses their representatives for government. Between elections, however, democracy is the process by which politicians rule the electorate. Tocqueville calls this a 'soft despotism' – the state is run by a tutelary power over which the people have little control; and elections simply provide the pretense of democratic form. "The only defense against this, Tocqueville thinks, is a vigorous political culture in which participation is valued, at several levels of government and in voluntary associations as well. But the atomism of the self-absorbed individual militates against this."[242] This is the fundamental paradox of democracy: the rhetoric of politicians celebrates the participation of the people in the political process (limited to voting) and yet asks them to sign away their liberty through periodic elections.

Democracy is premised on the realization of an informed and critically-thinking electorate. I think that it is obvious that our corporate media cannot be expected to properly inform the electorate, so we must rely on the critical thinking and 'political action' skills developed in our system of education … oh, oh. Communication and collaboration characterize the foundation of democratic decision-making. So, incongruously, we institute the secret vote and create the competitive process of electioneering to achieve these goals. How does the secret vote encourage debate and consensus seeking that is the hallmark of democracy? Instead, we vote on complex issues distilled to a simple binary (yes/no) choice on questions that presuppose the allowable answers. And how does the competitive process of choosing and electing candidates result in the most suitable person – the one who is best able to represent the constituency? The qualities of intellect and character that make good candidates (i.e., the competitive minion of the ruling elite) are not necessarily those that make good administrators for a democratic society.

We simplify community issues to caricatures, and we simplify governance to periodic elections. We discourage meaningful debate, critical thinking, and consensus by disengaging decision-making from public process. All things considered, a 'good citizen' is one who obeys the laws, pays taxes, votes ritualistically for pre-selected candidates, and otherwise minds his or her own business. The solution? Stop validating a process, or the façade of a process, that has nothing to do with democracy. In Saramago's novel *Seeing*, the population fails to go to the polls to elect their depot. The government is shaken by this event, but decides it was due to inclement weather on the day of voting. They stage another election, in which even fewer citizens participate. The leaders see this as a clever plot to overthrow the government and they institute marshal law, and send out undercover agents to discover the source of the insurrection. Voting is the assurance of a compliant public – not a forum for change.

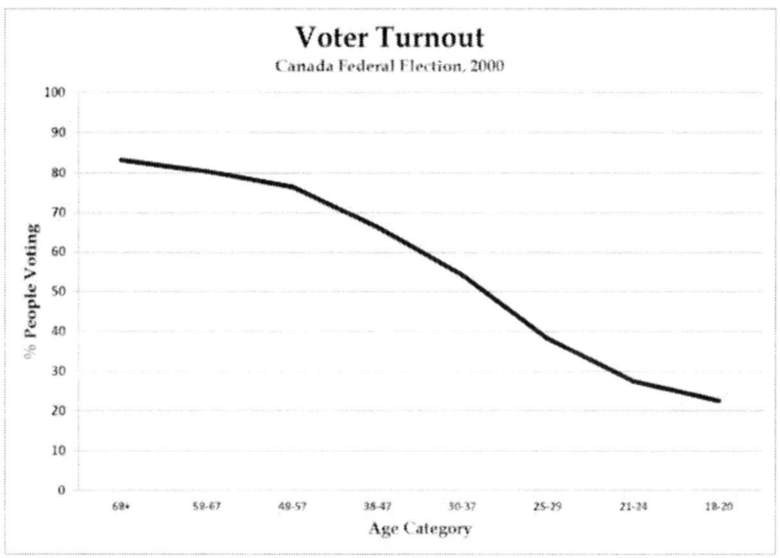

Figure 20 - Voter Turnout in Canada (by age)

And Saramago is not alone in thinking this. Voting turnout in Canada for federal elections has dropped from a high of almost 80% in the 1960s to about 60% in the 2000s[243]. This trend is mirrored in other countries without mandatory voting laws - another irony (Saramago's government should have thought of forcing people to vote). Youth voters in the 2000 federal election had a 20% voter turnout. Some reasons for this, according to the 2003 study, include negative public attitudes about politicians, government and candidates (53% of non-voters) and the meaninglessness of participation (25% of non-voters). Surprisingly, these percentages are the same when all respondents are included (voters and non-voters) – so it seems that many people voted out of some sort of ritualistic belief that the process of voting is a citizen's responsibility regardless of the fruitlessness of the act.

Clearly, there has been a steady decline in voter participation in the last half of a century, and fewer young people are voting. If the attitudes of the young voters don't change, a further erosion of average participation in elections can be anticipated as they move up the demographic chain. These trends have been met with an effort to heavily advertise before each election, selling the message that voting *is* democracy. It's not working, and the non-voters are staying away, not because they are shiftless and lazy, but because they simply understand that it is meaningless - which it is. Voting is "the technical debasing of views into mere preferences, of ideals into mere taste, of overall comprehension into quantification such that human aspirations and beliefs can be reduced to numerical digits."[244]

So, if democracy is not voting, what is? According to Berger, democracy "is a proposal (rarely realized) about decision-making; it has little to do with election campaigns. Its promise is that political decisions be made after, and in the light of, consultation with the governed. This is dependent upon the governed being adequately informed about the issues in question, and upon the decision-makers having the capacity and will to listen and take account of what they have heard. Democracy should not be confused with the 'freedom' of binary choices, the publication of opinion polls, or the crowding of people into statistics. These are its pretense."[245] Democracy is active participation – as opposed to the passive participation of sending representatives to serve their masters (and I don't mean their constituents) in the capitol. Democracy is discussion, it is expression, and it is inclusive. Graeber argues that in the existing system, elections are "where voting encourages one to reduce one's opponents' positions to a hostile caricature, or whatever it takes to defeat them" whereas "a consensus process is built on a principle of compromise and creativity, where one is constantly changing proposals around until one can come up with something everyone can at least live with. Therefore, the incentive is always to put the best possible construction on others' arguments." [246] Perhaps this could be measured by the participation of people in public discussion - through attendance at political forums or rallies, participation in discussion groups, writing a letter to the editor, calling into a talk show, or though petition, boycott and demonstration. On average, 35% of people do these things[247]. It would appear that the other half of the people who voted in the 2000 Canadian federal election did so out of some sort of ritual - not having been active in any other aspect of the democratic process.

In the context of this exploration of our ability to make change, it appears that voting in elections will not be of much assistance. Only 60 or 70% of people vote, and only about half of these people have been involved in the democratic process (beyond the punctuation at the end of the sentence). It is no wonder that politicians make very little effort to engage the citizenry in a debate about the issues - one just needs a majority of the few who vote to be allowed to rule for the next number of years. The truth about environmental issues, like climate change and water quality, and economic issues, like energy scarcity and mineral depletion, do not resonate with voters when offered a more pleasing siren's song. In fact, a politician would be foolish to engage in these issues in the current playing field of politics. You could be a politician who says, "we have some emerging challenges; it may require some lifestyle changes; it may cost some money", or you could be a politician who says, "there are no environmental issues as the extremist/alarmists would have you believe; keep enjoying yourselves and we'll deliver even more good things to your home once we are in power; we have to use our money to create jobs (and maybe a new curling rink)." The latter politician will indubitably get elected, it's a no-brainer: deny the facts, paint a rosy picture for the electorate, and when you are found out to have been wrong, you have already retired with a golden pension. As Thoreau said: 'All voting is a sort of gaming, like checkers or backgammon ...' When the shoe finally drops, people will be too busy adapting to the diminished environmental and social conditions than to spend much energy vilifying past politicians. Face it, the last people on earth who will lead us to a safer future will be the politicians. And most of the electorate already knows this.

And don't be fooled that the ruling elite wants anything other than the system as it currently functions (or doesn't function). In reality, participatory democracy is seen by them as a significant risk - they see democratic power in the filthy hands of the unwashed as a threat. People like Rousseau articulated this risk by presenting the actuality of parliamentary democracy: "The people of England regards itself as free; but it is grossly mistaken; it is free only during the election of members of parliament. As soon as they are elected, slavery overtakes it, and it is nothing." In other words: "Free election of masters does not abolish the masters or the slaves."[248] Too much democracy, real democracy, is a threat:

> Taking their cue from Joseph Schumpeter, many theorists have concluded that Rousseauvian democratic theory, with its emphasis on participatory "rule by the people," is unrealistic and impractical. Democracy, they argue, should be regarded as a quasi market where politicians compete to offer their leadership services to an electorate. The essence of democracy is held to be not the active deciding of issues by an electorate but an institutional arrangement for selecting political leaders by means of a competitive struggle for the people's votes. The value of democracy, it is said, does not lie in its enhancement of participatory autonomy. To the contrary, too much such autonomy, too much participation by the citizenry, can be destabilizing.[249]

As the adage goes: "If voting changed anything, they'd make it illegal."[250] In other words, you can vote, you just can't participate.

The Crisis of (Social) Capital

"If present sacrifices are substantial and future benefits collective, no action is possible. The bourgeois mentality, in other words, is so conditioned that it can never transcend the horizon of individual interests. When a given historical situation seems to call for such an effort, the response is recourse to rationalization which, while distorting reality, provides the needed justification for attitudes and actions which can pass the private interest test. This analysis explains one of the most obvious and yet puzzling things about capitalist society, why it can never act in advance to forestall a crisis, no matter how predictable it may be, but must always wait and act after the crisis has occurred."[251]

Paul Sweezy and Leo Huberman

Despite our promising beginnings (harnessing fire, developing language, inventing the wheel, publishing the book, and crafting the iPad) lately things appear to be going south. To be sure, I'm not blaming Apple for our current malaise. It could be just a coincidence that the last quarter century has experienced a sea change in the communication technologies that are designed solely, it seems, to support online dating that encourages our species to be fruitful and multiply (and, of course, fill the earth and subdue it). Seriously, though, what is the state of our social capital?

At the beginning of this century, Robert Putnam published a comprehensive study of the condition of what he termed 'social capital'. The consequences of social capital are described as mutual support, cooperation, trust, and institutional effectiveness. Putnam says:

> A society characterized by generalized reciprocity is more efficient than a distrustful society, for the same reason that money is more efficient than barter. If we don't have to balance every exchange instantly, we can get a lot more accomplished. Trustworthiness lubricates social life. Frequent interaction among a diverse set of people tends to produce a norm of generalized reciprocity. Civic engagement and social capital entail mutual obligation and responsibility for action. [252]

Unfortunately, each imaginable indicator of social capital has been in decline since the 1960s and this trend may have begun even earlier, right after World War II.

Putnam traces political participation, civic participation, religious participation, workplace connections, informal social connections, volunteerism, philanthropy, acts of altruism and reciprocity, and feelings of honesty and trust. Each has been in steady decline, particularly those activities that require cooperative behavior (as opposed to more individual 'expressive' behavior, like letter-writing or 'tertiary' membership that requires that the member only write a cheque each year). He traces the causes to pressures of time, inequality of income and wealth, urban sprawl, technology and mass media, among others. He argues that each of these contributes to declining social capital, though the most disturbing correlation was intergenerational - each generation since those born at the beginning of the 20th century has be less socially engaged. And he believes that "the biggest generational losses in engagement still lie ahead"[253]

Putnam argues that social capital improves the ability for citizens to resolve collective problems more readily. They provide social norms and networks that ensure compliance with the collectively desirable behavior. He also suggests that social capital is the foundation of trust which is fostered by repeated interactions with their fellow citizens both socially and through work relationships. Social capital also bridges the interests of all people by "widening our awareness of the many ways in which our fates are linked"[254]. Isn't it a bit disconcerting that as social capital declines, these positive values are being lost? That we are living more and more in a world fragmented by individual interests, where we are less and less able to resolve collective problems?

The impacts of what economists call 'transaction costs' which rise when people no longer trust each other, the costs incurred when community cooperation is replaced by corporate services, and the costs political extremism and the veritable Saturnalia of political corruption that result when people disengage from governance could alone bring our economies to their knees. When you add the other challenges to maintaining our way of life, the state of things is much more than dire.

In addition to the effects of electronic media, suburbanization, job insecurity and the rest of the list Putnam provides, there is the insidious impact of the loss of trust in government. Part of this loss of trust can be attributed to a general disengagement of citizens from participation, but part of it is well deserved. Take for example, the U.S. government under Bush the Second. During his tenure, the White House made every effort to occlude scientific opinion around global warming and its effects on climate change. James Hansen of the NASA Goddard Institute for Space Science has shared his story in *Storms of My Grandchildren: The Truth About the Coming Climate Catastrophe and our Last Chance to Save Humanity*. In it he describes the creation of a systemic ineptocracy in managing scientific process, a process in which public relations personnel redacted research from climate scientists to better align it to the ideology of the ruling regime and their corporate benefactors.

Another shocking example was the Senate testimony of Julie Gerberding on the public effects of climate change. Whole sections of the testimony related to extreme weather, air pollution, disease vectors, and food and water scarcity were removed from the testimony. The word-count went from over 3000 words to below 1500 words after the White House intervention. Some of the scientific conclusions that were redacted included the direct effects of weather on public well-being including the greater frequency and severity of heat waves and hurricanes, particularly as the U.S. demographic ages and moves to more vulnerable locations along the coast. Climate change would also have an effect on air quality by modifying weather patterns and pollutant concentrations due to higher surface temperatures and ozone formation. Studies show that some noxious weeds like ragweed and poison ivy grow faster under conditions of higher carbon dioxide concentrations and warm weather, which would exacerbate allergic reactions. The incidence of food- and water-borne diseases would be affected by changes in precipitation, temperature, and humidity. Other diseases like plague, Lyme disease, malaria, dengue fever, and West Nile virus are sensitive to seasonal patterns and climate changes would impact the range and incidence in the human population. Food scarcity issues due to changes in climate and other issues were also excluded from the testimony. In fact, every scientific concern about climate change on the well-being of the population was edited from the testimony - why?

The clearest reason is ideology. The U.S. government, among others who govern wealthy nations, do not accept any science that conflicts with their economic policy of growth and libertarian beliefs in 'personal responsibility'. To accept these inconvenient truths would oblige the government to act on behalf of the citizens, and it would mean a drastic reduction in fossil fuel emissions - which creates a politically difficult situation when there are no other energy sources currently available to replace them, as we have shown. What is more important than the crass irresponsibility of science denial, is the substantial loss of trust that this will eventually cause. When the governments of the world indeed begin to react in earnest to address the many converging challenges, where will be the public trust? And without public trust, how does a government affect a paradigm shift in public behavior?

The problem is that all science is vulnerable to attack, if there is motive. Science is based on controlled experiment and observation. A hypothesis is tested and the statistical data will either reject or fail-to-reject the hypothesis within a range of uncertainty. A typical research conclusion might sound something like 'the data fails-to-reject the hypothesis with a confidence of 9 out of 10 times.' Doesn't fail-to-reject mean that it proves it is correct? What about the other 1 out of 10 times? It's no wonder that the public believes that climate change is uncertain when there is actually a high level of scientific agreement on both the process and the impacts. What the scientist is saying above is that the data supports the hypothesis though there is a low statistical chance (10%) that the hypothesis should have been rejected. This does not necessarily mean that the hypothesis was wrong; it may mean that there was not enough data collected or that the methodology needs refinement. The more data that is collected, the more certain are the results; and as more scientists study the hypothesis, the more certain is the methodology. Complex systems like the metabolism of the whole world require a lot of scientists collecting a lot of data and working hard to improve the methodologies to eliminate causes of variance. There will always be statistical uncertainty in research, even though the conclusions become more certain. If you read the International Panel on Climate Change (IPCC) reports on the trends in warming, they use the terms of 'very high confidence' to suggest a degree of confidence of 9 out of 10. The IPCC's 2007 assessment states that there is a 'very high likelihood' that heat-trapping emissions from human activities have caused most of the globally observed increases in temperature since the middle of the twentieth century, and they give a likely range of average global temperature increase of 1.1°C to 6.4°C by the end of the 21st century. A 'Very High Likelihood'! This does not mean that global warming does not exist. The statistical uncertainty does not mean that the IPCC is uncertain. It means that it is happening, and that the impacts could range from bad to oh-so-unimaginably-bad. Why would governments deny the science? Why do people believe the absurdities spun by denialists? Is it because the principles of scientific research are too confusing? Why is the smoke from denialists so difficult to clear from media discourse once it has been emitted?

Chris Mooney argues that we are hard-wired to understand through our emotions, particularly when faced with a threat.[255] He quotes Arthur Lupia of the University of Michigan to suggests that a 'basic human survival skill' has been to resist threatening information while embracing friendly information – a 'fight-or-flight' reflex not only in the face of predators, but when exposed to information. The deliberation that follows the instantaneous emotion works within the setting that was created by the emotion. The subconscious negative response to a threatening piece of information guides the associations formed by the conscious mind: "Climate change is threatening, I don't believe it." New information being integrated into the schema of previously established beliefs will be challenged if it is inconsistent (and embraced if it aligns with these beliefs).

This 'disconfirmation bias', Mooney explains, encourages us refute views that we find uncongenial. And, of course, these refutations become more sophisticated as the denier becomes more sophisticated.

In the United States, the gap between the Democrats and the Republicans is 40% who accept the IPCC assertion that climate change is real and human caused: a study McCright and Dunlap[256] has shown that this gap has increased as scientific uncertainty has narrowed. A Democrat today is almost 70% likely to believe the effects of global warming have begun to happen, while a Republican is less than 30% likely. This can be compared to 60% and 50%, respectively, ten years ago. What would compel Republicans to become less convinced and Democrats more convinced as the science has become more certain? One result of the study suggests that "among elites and organizations within society, there is an enduring conflict between forces of reflexivity (those mostly on the Left who identify problems with our economic system) and forces of anti-reflexivity (those mostly on the Right who defend the industrial capitalist order of modernity against such critiques). In other words, issues like global warming and climate change threaten economic beliefs and result in ideological polarization between liberals and conservatives. Hamilton[257] supports these findings and adds that political polarization is greatest among those who are most confident that they understand the issue of climate change. Even with greater scientific understanding and near consensus, and even with more information available to the public, the sophisticated denier is less likely to change their opinion. A recent study by Kahan suggests that, for conservatives, more education correlated with more, not less, denial of the risks of climate change.

On the whole, the most scientifically literate and numerate subjects were slightly *less* likely, not more, to see climate change as a serious threat than the least scientifically literate and numerate ones. More importantly, greater scientific literacy and numeracy were associated with greater *cultural polarization*: respondents predisposed by their values to dismiss climate change evidence became more dismissive, and those predisposed by their values to credit such evidence became more concerned as science literacy and numeracy increased. It is suggested that this evidence reflects a conflict between two levels of rationality: the individual level, which is characterized by the citizens' effective use of their knowledge and reasoning capacities to form risk perceptions that express their cultural commitments; and the collective level, which is characterized by citizens' failure to converge on the best available scientific evidence on how to promote their common welfare.[258] Called the 'sophisticates effect', this shows that education and more scientific information, will have little impact on changing beliefs held by the ideologically conservative sector of society – the same sector that largely influences government policy.

The existing divide in the public perspective makes the necessary policy changes difficult, if not impossible, to implement in a timely way. As the effects of climate change become more evident, public ideology will likely become more conservative – the result of fear and uncertainty. This study indicates that a shift to the right will undermine the adoption of policy that might mitigate the worst effects – a positive feedback of ideological polarization delaying action which exacerbates the negative effects of climate change and causes public opinion to shift ever rightward amidst the hysterical effort to control and protect what is left over, thus magnifying the polarization and the effects.

Another conservative barrier to accepting the science of climate change beyond the defense of capitalism, is the defense of personal freedom. The premise is that different people have different goals, projects, and values, and that it is a good idea for the government to stay out of these choices. The government might be allowed to impose some limits, but it cannot decide what constitutes the good life. The threat of global warming, viewed in this light, becomes a threat to personal liberties as the government interferes with behavior and, worse, the market. The global collaboration that would be necessary to affect change is viewed as a world government that would threaten liberty everywhere. Government interference in the market also contradicts the ideology of the beatific invisible hand that distributes wealth most efficiently to the most deserving - if climate change were really a problem, the markets would solve it.

The Sucker Effect

"There ain't no causes - there ain't nothing but taking in this world, and he who takes the most is smartest - and it don't make a damn bit of difference *how*."[259]

Lorrain Hansberry

You can see that I have made every effort not to blame anyone for our predicament (except corporations, media, educational institutions, wealthy capitalists, politicians, and voters, that is). But now, perhaps, it's time to say exactly why 'we wouldn't do anything about it if we could'.

Basically, as Elinor Ostrom has said, "No one wants to be a 'sucker,' keeping a promise that everyone else is breaking."[260] Why in the world would anyone renounce a luxury while their neighbours refuse to do so - or worse, say they will, but sneak their luxuries when nobody is looking? Jim Hoggan says the same thing: "No one wants to give up their car, change their diet, or limit their consumption if their efforts will be rendered irrelevant by consumption patterns of those around them."[261] And why would anybody refuse to work to make something that is clearly in no one's best interest when they need to earn a living - to sell their labour-power: As Upton Sinclair has said, "It's difficult to get a man to understand something when his job depends on him not understanding it."

Joseph Heath relates some research about what he calls 'hyperbolic discounting'. In this research, people are offered a choice between an immediate gift or a better gift to be bestowed in the future. Most people took the immediate gift. When offered a gift in the future, or a better one a few days later, people chose the better gift. As the date approached, however, many people changed their minds to the lesser, but more immediate gift. Heath concludes that: "When the time to act is far away, we resolve to choose the greater good, but as the moment of decision approaches we give in to temptation and choose the lesser, more immediate good. … The whole problem with extreme hyperbolic discounting is that it makes people unwilling to tolerate short-term deprivations in order to achieve long-term benefits."[262]

Economically, too, it seems that when we have the money, we are too busy making money to address social change; and when we have the time, we have no money to make the necessary changes. Socially, when we try to use fright to affect change, we tend to paralyze people into inaction; but when we try to soothe we simply ensure inaction as business-as-usual. You've probably heard the story of the theatre fire. There was a full crowd at a theatre to watch *Waiting for Godot*, an uplifting story about purpose and human nature. Suddenly a person ran onto the stage yelling that there was a fire. People rose in fear and began moving towards the exits. A quick-witted person on stage saw that people were going to get hurt in the ensuing panic for the doors, so she went to the nearby piano on the side of the stage and began to play. The crowd immediately began to calm. She continued to play, and played so well, and with such assurance that people began to return to their seats … where they burned to death.

Conclusion

All I want is for future generations to just go,
'Fuck it. Had enough. ... Here's the truth.'

John Lydon

Now, doesn't that first chapter look a lot more appealing now? ... *we don't understand the problem; we couldn't do anything about it if we did; and we wouldn't do anything about it if we could.*

But what do you expect from a species that would plan a waterpark with surfing and river rafting in Arizona[263]? What do you expect from a species that would put a ski hill in Dubai[264]? What do you expect from a species that would cut down all the trees on an island they can't escape so as to simply roll statues around?

The bottom line is that we consume too much energy, this energy is finite and will be less available in the future, we need energy to make energy, we need energy for food, we need energy to do work at a rate that will sustain over 7 billion people, and we need energy to remove carbon dioxide from the atmosphere (or for geoengineering) to avoid the worst effects of climate change. If we fail to reduce energy consumption (which is pretty much a certainty) and if we fail to reduce the impact of greenhouse gases in the atmosphere, we should brace for a steady increase in average global temperatures with the concomitant extreme weather events, water scarcity, diminishing food productivity, not to mention migration and political unrest on a global scale. There are no technologies to fill the gap, and even if there were, the perversity of The Jevons Paradox in a capitalist economy will nullify any gains. Of course, by this time our economic system will have collapsed because of its inability to accumulate in a post-post-scarcity world. Politicians will be doing what they do best - making things worse, as they respond to (and exacerbate) public hysteria. As such, political cooperation and international investment to manage a transition will be non-existent.

Traumatized?

It is kind of a shame that we have collectively lacked the intelligence (that intelligence that we make so much fuss about, along with our opposable thumbs) to save our own skins. I was hoping that my children might enjoy their futures, and that their children would benefit from the fruits of human progress - that is, instead of fighting one another in the streets over the last scraps of civilization as we squeeze the last fucking drop of milk and honey from the miserable remains of God's nature.

The culmination of all of our challenges - our reluctance to face the realities of climate change; the declining availability of high EROEI energy; our careless consumption of non-renewable minerals; our mismanagement of our seas and our soil; our contamination of air, water and earth; our rapacious economic system premised on unlimited growth and waste; our inability to keep human populations under control; our unsustainable expectations of affluent lifestyles; our forfeiture of a functioning democratic decision-making process including education, media, and governance; our disengagement from our communities - will ultimately overwhelm us.

We are already extinct.

"When we speak today of the world ecological crisis however, we are referring to something that could turn out to be *final*, that is, there is a high probability, if we do not quickly change course, of a *terminal crisis* – a death of the whole period of human dominance of the planet. Human actions are generating environmental changes that threaten the extermination of most species on Earth, along with civilization, and conceivably our own species as well."[265]

John Bellamy Foster

"With a kind of savage justice, climate change is an issue which exposes the weakest link in the cultural mindset of Western market capitalism: the collective capacity for self-restraint in pursuit of a common good."[266]

Madeleine Bunting

"Fifteen billion years of evolution have produced a being able to unearth the origins of the universe that gave it life, able to decode the behavior of atoms and galaxies, to explore the solar system, to master the forces of nature, but unable to marshal its resources in such a way as to avert its own elimination."[267]

Hubert Reeves

"Experts in a range of fields have begun to see the same closing door of opportunity, begun to warn that these years may be the last when civilization still has the wealth and political cohesion to steer itself towards caution, conservation, and social justice. … more than half the world's Nobel laureates warned that we might have only a decade or so left to make our system sustainable. … Pentagon predicts worldwide famine, anarchy, and warfare 'within a generation' should climate change fulfill the more severe predictions. … Martin Rees [says] 'The odds are no better than fifty-fifty that our present civilization … will survive to the end of the present century … unless all nations adopt a low-risk and sustainable policies based on present technology.'"[268]

Ronald Wright

"Half the world's tropical and temperate forests are now gone. The rate of deforestation in the tropics continues at about an acre a second. About half the wetlands and a third of the mangroves are gone. An estimated 90 percent of the large predator fish are gone, and 75 percent of marine fisheries are now overfished or fished to capacity. Twenty percent of the corals are gone, and another 20 percent severely threatened. Species are disappearing at rates about a thousand times faster than normal. The planet has not seen such a spasm of extinction in sixty-five million years, since the dinosaurs disappeared. Over half the agriculture land in drier regions suffers from some degree of deterioration and desertification. Persistent toxic chemicals can now be found by the dozens in essentially each and every one of us.

Human impacts are now large relative to natural systems. The earth's stratospheric ozone layer was severely depleted before the change was discovered. Human activities have pushed atmospheric carbon dioxide up by more than a third and have started in earnest the dangerous process of warming the planet and disrupting climate. Everywhere earth's ice fields are melting. Industrial processes are fixing nitrogen, making it biologically active, at a rate equal to nature's; one result is the development of more than two hundred dead zones in the oceans due to overfertilization. Human actions already consume or destroy each year about 40 percent of nature's photosynthetic output, leaving too little for other species. Freshwater withdrawals doubled globally between 1960 and 2000, and are now over half of accessible runoff. The following rivers no longer reach the oceans in the dry season: the Colorado, Yellow, Ganges, and Nile, among others."[269]

James Gustave Speth

"Citizen movements see a very different reality. Focused on people and the environment, they see the world in a crisis of such magnitude that it threatens the fabric of civilization and the survival of the species – a world of rapidly growing inequality, erosion of relationships of trust and caring, and failing planetary life support systems…."[270]

Alternatives to Economic Globalization

"Even if we cut emissions by 60 per cent to 12 gigatons a year it wouldn't be enough. ... but did you know that the exhalations of breath and other gaseous emissions by the nearly 7 billion people of Earth, their pets and their livestock are responsible for 23 per cent of all greenhouse gas emissions? If you add on the fossil fuel burnt in the total activity of growing, gathering, selling and serving food, all of this adds up to about half of all carbon dioxide emissions. ... I do not see how the 60 per cent reduction can be achieved without a great loss of life."[271]

James Lovelock

"For example, there is the staggering (and largely unremarked) loss of topsoil: more than 25 million acres degraded or lost annually. As Jared Diamond has said, 'There are about a dozen major environmental problems, all of them sufficiently serious that if we solved eleven of them and didn't solve the twelfth, whatever that twelfth is, any could potentially do us in.'"[272]

Patrick Curry

"If a society develops a serious ingenuity gap - that is, if it loses the race between requirement and supply - prosperity falls in the regions already affected by scarcity, and people usually migrate out of those regions in large number. Social dissatisfaction rises, especially among marginal groups in ecologically fragile rural areas and urban squatter settlements. These changes undermine the government's legitimacy and raise the likelihood of widespread and chronic civil violence. Violence further erodes the society's capacity to supply ingenuity, especially by causing human and financial capital to flee. Such societies risk entering a downward and self-reinforcing spiral of crisis and decay."[273]

Thomas Homer-Dixon

"It is not a question of our encountering the crisis and resolving it through technology. The crisis is not simply something we can examine and resolve. We *are* the environmental crisis. … The environmental crisis is inherent in everything we believe and do; it is inherent in the context of our lives."[274]

Neil Evernden

"The primary challenge facing the people of the world is, literally, survival."[275]

Noam Chomsky

Appendix: Figures and Figuring

World Primary Energy Consumption (Annually):
470 exaJoules
470,000,000,000,000,000,000 Joules

Barrel of Oil Equivalent (BOE) energy:
6.1 gigaJoules / BOE
6,100,000,000 Joules/BOE

Barrel of Oil volume:
42 usgallons
159 liters

Volume of Empire State Building:
37 million cubic feet

Continuous Power from a Human:
100 W (100 J/s)

Calculation of Energy-People:
Energy-People
= Primary Energy Consumed Annually x 30% conversion efficiency / Annual work per person
= 470 exaJoules x 0.3 / (100 J/s x 3600 s/hr x 8 hr/day x 365 d/yr)
= 136 billion people (or, with 7 billion people on the planet, 20 Energy-People each)

Global Electricity Produced Annually:
20,000,000,000,000 kWh

% of Global primary energy converted to electricity:
15.3%

Global PV Capacity
21,300,000 kW
@ 1020 kWh* annual production / kW installed on average globally
(*this value depends upon location, operation, maintenance, weather, etc.)

Global PV Electricity Produced Annually:
21,500,000,000 kWh

What is a kWh?
3,600,000 J / kWh

Global PV Electricity Produced Annually:
77,400,000,000,000,000 J

% PV in Global Electricity Production
= 21,500 GWh / 20,000,000,GWh = 0.11%
By 2020, the global electricity demand is expected to be 25,000,000,000,000 kWh
By 2020, the PV capacity may reach 80,000 MW @ 1020 kWh/kW = 81,600,000,000 kWh
By 2020, **% PV in Global Electricity Production**
= 81,600 GWh / 25,000,000,GWh = 0.33%

Transportation Fuels (**2008 Stats Canada**):

Diesel Consumption Freight (petajoule, 15 zeros) = 785.4 PJ
Diesel Consumption Passenger (petajoule, 15 zeros) = 58.1 PJ

Heating Value Diesel
38.6 MJ/liter

Diesel Consumption
= (785.4 + 58.1) x 10^9 MJ/year / 38.6 MJ/liter / 365 d/year
= 60,000,000 liters/day

Gasoline Consumption Passenger Vehicles = 1078.5 PJ
Gasoline Consumption Freight = 232.3 PJ
Heating Value Gasoline = 34.8 MJ/liter

Gasoline Consumption
= (1078.5 + 232.3) x 10^9 MJ/year / 34.8 MJ/liter / 365 d/year
= 103,200,000 liters/day

Canada Arable Land = 45.1 million hectares
Land under corn cultivation = 1.2 million hectares
Land under what cultivation = 8.6 million hectares

Corn to Ethanol Yield = 2500 liters / hectare
Wheat to Ethanol Yield = 1000 liters / hectare
Heating value of Ethanol = 29.8 MJ/kg
Density of Ethanol = 0.789 kg/l

Total ethanol production from current corn production
= 1.2 million ha x 2500 liters/ha/y / 365 d/y
= 8.2 million liters / d

Total ethanol production from current wheat production
= 8.6 million ha x 1000 liters/ha/y / 365 d/y
= 23.5 million liters / d

Heating value of total ethanol from grains
= (8.2 + 23.5) million liters x 29.8 MJ/kg x 0.789 kg / l
= 272 PJ / year
= 272 PJ / (1078.5 + 232.3) PJ = 20.8% of total energy demand for gasoline.

If all of the arable land in Canada was put into wheat production:

= 45.1 million ha x 1000 liters/ha/y / 365 d/y
= 123.6 million liters / d
= 123.6 million liters x 29.8 MJ/kg x 0.789 kg / l
= 1060 PJ (or 81% of demand)

World Arable Land:
1400 million hectares

Assuming an average ethanol yield:
3000 liters / ha

Global production
= 1,400,000,000 ha x 3000 liters/ha
= 4,200 billion liters/year
= 4200 billion l/y x 24 MJ/l
= 100 EJ / year
= 100 EJ/year production / 475 EJ/year consumption
= 21.2 %

Waste-to-Energy:
North American production of waste:
2 to 3 kg/day/person
Household waste (Canada 2004)[276]
13.4 million tonnes
Average Canadian production of waste:
418 kg/person/year
Average diversion to recycling (Canada, 2004)
27% by weight
Heating Value of MSW (including recyclables)[277]
11 million Btu/ton (12 GJ/tonne)
Heating value of MSW (without recyclables)
6 to 9 GJ/tonne

At a 30% conversion rate
= 6 GJ/tonne x 0.3 / 3600 s/hour
= 500 kWh/tonne

Electricity from waste
= 13.4 million tonnes/year x 500 kWh/tonne
= 6.7 billion kWh/year
Canadian electricity consumption[278]
500 billion kWh
Residential electricity consumption
= about 30%
Percentage of waste to energy
= 6.7 billion kWh/year / 500 billion kWh/year
= 1.3% of total electricity consumption
= 6.7 billion kWh/year / 30% of 500 billion kWh/year
= 4.3% of household electricity

Total energy consumed in Canada
7600 PJ or 7600 million GJ/year

Waste heating value
= 6 GJ/tonne x 13.4 million tonnes/year
= 80.4 million GJ/year

Percentage waste to energy
=80.4 million GJ/y / 7600 million GJ/year
= 1%

Acknowledgements

The efforts of the Fescue Collective must be acknowledged, as they strive for a level of excellence already achieved by publishers like Monthly Review Press and New Society Publishers. I would also like to thank the critical thinkers who have greatly informed the writing of this book: John Bellamy Foster and the writers for *Monthly Review* magazine; Herman Daly; Annie Leonard, Richard Heinberg, James Gustave Speth, Neil Evernden, Thomas Homer-Dixon, Ted Trainer, Patrick Curry, David Suzuki, and countless others. (Of course, despite their clear thinking and presentation, all errors, misconceptions and opinions in this work remain my own.)

And where would we be without the courageous scientists that face political pressure and targeted censure for simply conducting and publishing quality, peer-reviewed science for the greater good? Professors like Michael Mann, Andrew Weaver, and Dan Johnson come to mind. And I would especially like to thank James Hansen who took an inspirational turn away from the conservative objectivity of science to face the issues on a personal level by writing *Storms of My Grandchildren*.

And, finally, this is for my loving and supportive wife, always, and our two children for whom I fear, as the coming crises materialize. For the love of humanity, I wish it were different.

Endnotes

1 Slavoj Žižek, *For they know not what they do: Enjoyment as a political factor* (New York: Verso, 2008), 53.

2 Slavoj Žižek, *For they know not what they do: Enjoyment as a political factor* (New York: Verso, 2008), 131.

3 Slavoj Žižek, *For they know not what they do: Enjoyment as a political factor* (New York: Verso, 2008), 9.

4 Slavoj Žižek, *The Ticklish Subject* (New York: Verso, 1999), 348.

5 Frederick Kirschenmann, "Do increased energy costs offer opportunities for a new agriculture?" in Fred Magdoff and Brian Tokar, eds., *Agriculture and food in crisis: conflict, resistance, and renewal* (New York: Monthly Review Press, 2010), 234.

6 Patrick Curry, *Ecological ethics: An introduction* (Cambridge: Polity Press, 2006).

7 Erik Assadourian, "The rise and fall of consumer cultures," in Linda Starke & Lisa Mastny. eds., *State of the World, 2010: Transforming Cultures* (New York: W.W. Norton & Company, 2010), 4.

8 Edward O. Wilson, *The future of life* (New York: Vintage Books, 2002), 23.

9 Annie Leonard, *The story of stuff: How our obsession with stuff is trashing the planet, our communities, and our health – and a vision for change* (Toronto: Free Press. 2010), 178.

10 Ronald Inglehart and Hans-Dieter Klingemann, *Genes, culture, democracy, and happiness* (2000), http://www2000.wzb.eu/alt/iw/pdf/genecult.pdf

11 Victor Lebow, as quoted in Annie Leonard, *The story of stuff: How our obsession with stuff is trashing the planet, our communities, and our health – and a vision for change* (Toronto: Free Press. 2010), 160.

12 Thorstein Veblen, *The theory of the leisure class* (New York: Penguin Books, 1899, 1967), 69.

13 Kai N. Lee, "An Urbanizing World," in Linda Starke, ed., *State of the World 2007: Our Urban Future* (New York: W.W. Norton & Company, 2010), 6.

---

14 British Petroleum. *Statistical review*, http://www.bp.com.
15 W.H. Ziegler, C.J. Campbell, and J.J. Zagar, "Peak oil and gas," *Swiss Bulletin for Applied Geology* 14 (2009).
16 Doug Horner, "What lies beneath: Almost 100,000 spent oil and gas wells litter Alberta. Who will pay the cleanup cost? ," *Alberta Views* (2012), http://www.albertaviews.ab.ca/2011/03/01/what-lies-beneath/
17 Thomas King, *The truth about stories: A native narrative* (Toronto: Anansi, 2003), 163.
18 From *The Ballad of Jed Campett* by Paul Henning, and better known as the theme song for the television show, *The Beverly Hillbillies*.
19 Eswaran Subrahmanian and V.S. Arunachalam, *Energy and material scarcity: Achilles heel of global manufacturing* (2010), Center for Science, Technology and Policy, http://www.cstep.in/node/254
20 H. Wouters, and D. Bol, *Material Scarcity* (The Netherlands: Materials Innovation Institute, 2009).
21 Chart is based on presentation by Subrahmanian & Arunachalam (see endnote 19).
22 Johan Rockström, et. al., "Planetary boundaries: Exploring the safe operating space for humanity," *Ecology and Society* 14 (2009), http://www.ecologyandsociety.org/vol14/iss2/art32/
23 John Bellamy Foster, Brett Clark and Richard York, *The ecological rift: Capitalism's war on the earth* (New York: Monthly Review Press, 2010), 14.
24 Patrick Déry, *Pérenniser l'agriculture* (2007), http://www.caaaq.gouv.qc.ca/userfiles/File/MEMOIRE(1)/02-07-Saguenay-Dery,Patrick.pdf
25 Maude Barlow and Tony Clarke, *Blue gold: The battle against corporate theft of the world's water* (Toronto: Stoddart Publishing Co. Ltd., 2002), 5.
26 Maude Barlow, *The global water crisis and the coming battle to the right of water* (Toronto: McClelland & Stewart Ltd., 2007), 3.
27 Bob Willard, *The NEXT sustainability wave* (Gabriola Island: New Society Publishers, 2005), 96.

[28] Maude Barlow and Tony Clarke, *Blue gold: The battle against corporate theft of the world's water* (Toronto: Stoddart Publishing Co. Ltd., 2002), 32.

[29] Fred Pearce, *When the rivers run dry: Journeys into the heart of the world's water crisis* (Toronto: Key Porter Books Limited, 2006), 382.

[30] James Gustave Speth, *The bridge at the edge of the world: Capitalism, the environment, and crossing from crisis to sustainability* (New Haven: Yale University Press, 2008), 1.

[31] Cormac Cullinan, *Wild law: A manifesto for earth justice*, second edition (Vermont: Chelsea Green Publishing, 2011), 40.

[32] Michael P. Wilson and Megan R. Schwarzman, "Toward a new U.S. chemicals policy: Rebuilding the foundation to advance new science, green chemistry, and environmental health," *Environmental Health Perspectives* 117 (August 2009).

[33] Gary Gardner and Terry Prugh, "Seeding the Sustainable Economy," in Linda Starke, ed., *State of the World 2008: Innovations for a Sustainable Economy* (New York: W.W. Norton & Company, 2008), 12.

[34] Richard Heinberg, *PowerDown: Options and actions for a post-carbon world* (Gabriola Island: New Society Publishers, 2004), 135.

[35] David Orr, as quoted in David Brower, *Let the mountains talk, let the rivers run: A call to those who would save the earth* (Gabriola Island: New Society Publishers, 2000), 93.

[36] An extensive database on the denial industry may be found on the website DESMOGBLOG.COM, http://www.desmogblog.com/

[37] The following discussion is based on a clear summary presented in: John W. Farley, "The scientific case for modern anthropogenic global warming," *Monthly Review* 60 (July-August 2008): 68-90.

[38] National Oceanic and Atmospheric Administration, *Coral reefs - An important part of our future,* http://www.noaa.gov/features/economic_0708/coralreefs.html (retrieved June 2012)

[39] GwynneDyer, *Climate wars* (Toronto: Random House Canada, 2008), 92.

---

40 GwynneDyer, *Climate wars* (Toronto: Random House Canada, 2008), 5.

41 Nicholas Stern, "The Economics of climate change," in S Stephen M. Gardiner, Simon Caney, Dale Jamieson, and Henry Shue, eds., *Climate ethics: Essential readings* (New York: Oxford University Press, 2010), 45.

42 Linda Starke, ed., *Vital signs 2011: The trends that are shaping our future* (New York: W.W. Norton & Company, 2011), 29.

43 Prof. Geoff Hammond and Craig Jones, *Inventory of carbon & energy (ICE)*, Version 1.6a (2008), http://perigordvacance.typepad.com/files/inventoryofcarbonandenergy.pdf

44 Anna Stoppato, "Life-cycle assessment of photovoltaic electricity generation," *Energy* 33 (February 2008): 224-232.

45 A.F. Sherwani, J.A. Usmani, and Varun, "Life cycle assessment of solar PV based electricity generation systems: A review," *Renewable and Sustainable Energy Reviews* 14 (January 2010): 540-544.

46 Varun, I.K. Baht, and Ravi Prakash, "LCA of renewable energy for electricity generation systems: A review," *Renewable and Sustainable Energy Reviews* 13 (June 2009): 1067-1073.

47 Yingling Liu, "The dirty side of a 'green' industry," Worldwatch Institute, March 14, 2008, http://www.worldwatch.org/node/5650.

48 Varun, I.K. Baht, and Ravi Prakash, "LCA of renewable energy for electricity generation systems: A review," *Renewable and Sustainable Energy Reviews* 13 (June 2009): 1067-1073.

49 Mark Jacobson, "A path to sustainable energy by 2030," *Scientific American* (November 2009): 58-65, http://www.stanford.edu/group/efmh/jacobson/Articles/I/sad1109Jaco5p.indd.pdf

50 Linda Starke, ed., *Vital signs 2011: The trends that are shaping our future* (New York: W.W. Norton & Company, 2011).

---

51 Erik Assadourian, "The rise and fall of consumer cultures," in Linda Starke & Lisa Mastny. eds., *State of the World 2010: Transforming Cultures* (New York: W.W. Norton & Company, 2010), 7.

52 Worldwatch Institute, *Biofuels for transportation: Global potential and implications for sustainable agriculture and energy in the 21st Century* (2006), http://www.worldwatch.org/system/files/EBF038.pdf

53 Douglas Bradley, *Canada report on bioenergy 2008*, http://www.bioenergytrade.org/downloads/canadacountryreportjun2008.pdf

54 Asia-Pacific Economic Cooperation (APEC), *Survey of biomass resource assessments and assessment capabilities in APEC countries* (2008), http://www.biofuels.apec.org/pdfs/ewg_2008_biomass_resource_assessment.pdf

55 Godfrey Boyle, *Renewable energy: Power for a sustainable future*, 2nd ed. (Cambridge: University Press, 2004).

56 (S&T)² Consultants Inc., *The addition of ethanol from wheat to GHGenius* (2003), http://www.manitoba.ca/iem/energy/biofuels/ethanol/files/wheat_ethanolreport.pdf

57 John Michael Greer, *The long descent: A user's guide to the end of the industrial age* (Gabriola Island: New Society Publishers, 2008), 17.

58 Brian Tokar, "Biofuels and the global food crisis, in Fred Magdoff and Brian Tokar, eds., *Agriculture and food in crisis: conflict, resistance, and renewal* (New York: Monthly Review Press, 2010), 125.

59 David Pimentel and Tad W. Patzek, "Ethanol production using corn, switchgrass, and wood; biodiesel production using soybean and sunflower," *Natural Resources Research* 14 (March 2005): 65-76. http://www.c4aqe.org/Economics_of_Ethanol/ethanol.2005.pdf

60 Canadian Renewable Fuels Association, *Growing beyond oil delivering our energy future: A report card on the Canadian renewable fuels industry* (2010), http://www.greenfuels.org/uploads/documents/crfareportcardenglish2010final.pdf

61 United Stated Department of Agriculture (USDA). *Biofuels Canada 2007*, http://www.fas.usda.gov/gainfiles/200708/146292128.pdf

62 Climate Change Solutions, *Canada biomass-bioenergy report* (2006), http://www.bioenergytrade.org/downloads/canadacountryreportmay312006.pdf

63 B.E. Holbein, J.D. Stephen, and D.B. Layzell, D.B., *Canadian biodiesel initiative: Aligning research needs and priorities with the emerging industry* (2004), http://www.iseee.ca/media/uploads/documents/BIOCAP/BIOCAP_Biodiesel04_Final.pdf

64 Jane Earley, and Alice McKeown, *Smart Choices for Biofuels* (2009), http://www.worldwatch.org/files/pdf/biofuels.pdf

65 Worldwatch Institute, *Biofuels for transportation: Global potential and implications for sustainable agriculture and energy in the 21st Century* (2006), http://www.worldwatch.org/system/files/EBF008_1.pdf

66 William T.W. Woodward, *The potential of alfalfa, switchgrass, and miscanthus as biofuel crops in Washington* (2008), http://www.wa-hay.org/Proceedings/08%20Proceedings/Potential%20Biofule%20Crops%20-%20Woodward.pdf

67 Jason Hill, Erik Nelson, David Tilman, Stephen Polsaky, and Douglas Tiffany, *Environmental, economic, and energetic costs and benefits of biodiesel and ethanol biofuels* (2006), http://cedarcreek.umn.edu/hilletal2006.pdf

68 Jane Earley, and Alice McKeown, *Smart Choices for Biofuels* (2009), http://www.worldwatch.org/files/pdf/biofuels.pdf

69 ICLEI Global., *The biodiesel alternative* (2007), http://www.iclei.org/fileadmin/user_upload/documents/ANZ/CCP/CCP-AU/Projects/Biodiesel/Chapter2_0710_update.pdf

70 Brian Tokar, "Biofuels and the global food crisis, in Fred Magdoff and Brian Tokar, eds., *Agriculture and food in crisis: conflict, resistance, and renewal* (New York: Monthly Review Press, 2010), 127.

71 Joseph Fargione, Jason Hill, David Tilman, Stephen Polasky, and Peter Hawthorne, "Land clearing and biofuel carbon debt," *Science* 29 (February 2008), http://www.sciencemag.org/cgi/content/abstract/1152747v1

---

72 Brian Tokar, "Biofuels and the global food crisis, in Fred Magdoff and Brian Tokar, eds., *Agriculture and food in crisis: conflict, resistance, and renewal* (New York: Monthly Review Press, 2010), 131.

73 J.A. Mathews, "Biofuels: What a biopact between North and South could achieve," *Energy Policy 35* (2007): 3550-3570.

74 Jean Bricmont, *Humanitarian Imperialism: Using human rights to sell war* (New York: Monthly Review Press, 2006).

75 Brian Tokar, "Biofuels and the global food crisis, in Fred Magdoff and Brian Tokar, eds., *Agriculture and food in crisis: conflict, resistance, and renewal* (New York: Monthly Review Press, 2010), 125.

76 Linda Starke, ed., *Vital signs 2011: The trends that are shaping our future* (New York: W.W. Norton & Company, 2011).

77 John Michael Greer, *The long descent: A user's guide to the end of the industrial age* (Gabriola Island: New Society Publishers, 2008), 110.

78 James Hoggan, *Climate cover-up: The crusade to deny global warming* (Vancouver: Greystone Books, 2009), 194.

79 Jan Willem Storm Van Leeuwan and Philip Smith, *Can nuclear power provide energy for the future; would it solve the CO2 emission problem?* (2002), http://greatchange.org/bb-thermochemical-nuclear_sustainability_rev.pdf

80 Jim Harding, *Canada's deadly secret: Saskatchewan uranium and the global nuclear system* (Halifax: Fernwood Publishing, 2007).

81 Thomas Homer-Dixon, "Our Fukishima moment," *The Globe & Mail*, March 18, 2011, http://www.theglobeandmail.com/news/opinions/opinion/our-fukushima-moment/article1946562/

82 Amory Lovins, as quoted in Gwynne Dyer, *Climate wars* (Toronto: Random House Canada, 2008), 186.

83 "Coal-fired power plants capacity to grow by 35 percent in next 10 years," EngineerLive, November 8, 2012, http://www.engineerlive.com/Power-Engineer/Focus_on_Coal/Coal-fired_power_plants_capacity_to_grow_by_35_per_cent_in_next_10_years/21600/

[84] Thomas Homer-Dixon, "Our Fukishima moment," *The Globe & Mail*, March 18, 2011, http://www.theglobeandmail.com/news/opinions/opinion/our-fukushima-moment/article1946562/

[85] Karl Marx, as quoted in Michael A. Lebowitz, *Beyond capital: Marx's political economy of the working class*, second ed. (New York: Palgrave MacMillan, 2003), 111.

[86] William Freudenburg, "Polluters' shell game: The worst polluters are hiding their miniscule contribution to the economy," Worldwatch 22 (January 2009): 17-21.

[87] James Gustave Speth, *The bridge at the edge of the world: Capitalism, the environment, and crossing from crisis to sustainability* (New Haven: Yale University Press, 2008), 56.

[88] William Freudenburg, "Polluters' shell game: The worst polluters are hiding their miniscule contribution to the economy," Worldwatch 22 (January 2009): 17-21.

[89] John Bellamy Foster, Brett Clark and Richard York, *The ecological rift: Capitalism's war on the earth* (New York: Monthly Review Press, 2010), 383.

[90] John Bellamy Foster, Brett Clark and Richard York, *The ecological rift: Capitalism's war on the earth* (New York: Monthly Review Press, 2010), 181.

[91] Herman E. Daly, *Beyond growth* (Boston: Beacon Press, 1996), 36.

[92] James Gustave Speth, *The bridge at the edge of the world: Capitalism, the environment, and crossing from crisis to sustainability* (New Haven: Yale University Press, 2008), 149.

[93] James Gustave Speth, *The bridge at the edge of the world: Capitalism, the environment, and crossing from crisis to sustainability* (New Haven: Yale University Press, 2008), 41.

[94] James Lovelock, *The vanishing face of Gaia: A final warning* (New York: Penguin Books, 2009), 50.

[95] Hsien Khoo, "Life cycle impact assessment of various waste conversion technologies," *Waste Management* 29 (June 2009): 1982-1900.

---

[96] C.S. Psomopolous, A. Bourka, and N.J. Themelis "Waste-to-energy: A review of the status and benefits in USA," *Waste Management* 29 (2009): 1718-1724.

[97] Human Resources and Skills Canada, *Indicators of well-being in Canada*, http://www4.hrsdc.gc.ca/.3ndic.1t.4r@-eng.jsp?iid=37

[98] Giovanni De Feo, and Carmela Malvano, "The use of LCA in selecting the best MSW management system," *Waste Management* 20 (June 2009): 1901-1915.

[99] Göran Finnveden, Jessica Johansson, Per Lind, and Åsa Moberg, *Life Cycle Assessments of energy from solid waste* (2000), http://www.imamu.edu.sa/topics/IT/IT%206/Life%20Cycle%20Assessments%20of%20Energy%20from%20Solid%20Waste.pdf Richard A. Dennison, *Environmental life-cycle comparisons of recycling, landfilling, and incineration: A review of recent studies* (1996), http://www.edf.org/sites/default/files/1340_denison.pdf

[100] Braum Barber and Leona Rousseau, "The Living Home: Building it into the curriculum," in Lucas Johnston, ed., *Higher education for sustainability: Cases, challenges, and opportunities from across the curriculum* (New York: Routledge, 2012).

[101] Intergovernmental Panel on Climate Change, *IPCC Special Report on Renewable Energy Sources and Climate Change Mitigation* (2011), http://www.ipcc-wg3.de/publications/special-reports/srren [Emphasis added]

[102] Erik Assadourian, "The rise and fall of consumer cultures," in Linda Starke & Lisa Mastny. eds., *State of the World 2010: Transforming Cultures* (New York: W.W. Norton & Company, 2010), 7.

[103] John Bellamy Foster, *Ecology against capitalism* (New York: Monthly Review Press, 2002), 94.

[104] Stanley Jevons, as quoted in John Bellamy Foster, Brett Clark and Richard York, *The ecological rift: Capitalism's war on the earth* (New York: Monthly Review Press, 2010), 171.

[105] Paul Mobbs, *Energy beyond oil* (Leicester, UK: Troubadour Publishing Ltd, 2005), 32.

106 John Bellamy Foster, Brett Clark and Richard York, *The ecological rift: Capitalism's war on the earth* (New York: Monthly Review Press, 2010), 189.

107 Anna Lappé, "The climate crisis on our plates," in Linda Starke, ed., State of the World 2011: Innovations that nourish the planet (New York: W.W. Norton & Company, 2011).

108 Minqi Li, "An age in transition: The United States, China, peak oil and the demise of neoliberalism," *Monthly Review* 59 (April 2008), 32.

109 Jess Jenkins, Ted Nordhaus, and Michael Shellenberger, *Energy emergence: Rebound & backfire as emergent phenomena* (2011), http://thebreakthrough.org/blog/Energy_Emergence.pdf

110 Jess Jenkins, Ted Nordhaus, and Michael Shellenberger, *Energy emergence: Rebound & backfire as emergent phenomena* (2011), http://thebreakthrough.org/blog/Energy_Emergence.pdf

111 Annamaria Lusardi, Daniel Schneider, and Peter Tufano, *Financially fragile households: Evidence and Implications* (2011), http://www.brookings.edu/~/media/Files/Programs/ES/BPEA/2011_spring_bpea_papers/2011_spring_bpea_conference_lusardi.pdf

112 Michael I. Norton, and Dan Ariely, "Building a better America – One wealth quintile at a time," *Perspectives on Psychological Science* 6:9, http://www.people.hbs.edu/mnorton/norton%20ariely%20in%20press.pdf

113 Al Gore, *The assault on reason* (New York: The Penguin Press, 2007), 74.

114 Lawrence Mishel, and Josh Bivens, "Occupy Wall Streeters are right about skewed economic rewards in the United States," Economic Policy Institute, October 26, 2011, http://www.epi.org/publication/bp331-occupy-wall-street/

115 David Harvey, *A brief history of neoliberalism* (New York: Oxford University Press, 2007).

---

[116] John Bellamy Foster, and Robert W. McChesney. "The global stagnation and China," *Monthly Review* 63 (February 2012), http://monthlyreview.org/2012/02/01/the-global-stagnation-and-china

[117] Naomi Klein, *The shock doctrine: The rise of disaster capitalism* (Toronto: Alfred A. Knopf Canada, 2007).

[118] Minqi Li, "An age in transition: The United States, China, peak oil and the demise of neoliberalism," *Monthly Review* 59 (April 2008), 23.

[119] Prime Minister Stephen Harper, as quoted in "that's that," *this Magazine*, March/April 2012. Emphasis added.

[120] Robert D. Putnam, *Bowling Alone* (New York: Simon & Schuster, 2001).

[121] David McNally, "Slump, austerity and resistance," in Leo Panich, Greg Albo, and Vivek Chibber, eds., *The Crisis and the Left: The Socialist Register 2012* (New York: The Monthly Review Press, 2011): 36-63, 41.

[122] Naomi Klein, *The shock doctrine: The rise of disaster capitalism* (Toronto: Alfred A. Knopf Canada, 2007), 513.

[123] David Harvey, *The enigma of capital and the crises of capitalism.* (London: Profile Books, 2010), 59.

[124] David Harvey, *A brief history of neoliberalism* (New York: Oxford University Press, 2007), 185.

[125] Michael Renner, "Broadening the understanding of security," in Linda Starke & Lisa Mastny. eds., *State of the World 2010: Transforming Cultures* (New York: W.W. Norton & Company, 2010), 127-132.

[126] Annie Leonard, *The story of stuff: How our obsession with stuff is trashing the planet, our communities, and our health – and a vision for change* (Toronto: Free Press. 2010), 150.

[127] John Kenneth Galbraith, *The culture of contentment* (New York: Houghton Mifflin Company, 1992), 135.

[128] John Kenneth Galbraith, *The Great Crash: 1929* (New York: Houghton Mifflin Company, 1954), 76.

[129] R.H. Tawney, *The acquisitive society* (New York: Dover Publications, Inc., 1920), 30.

130 Istvan Mészáros, *The structural crisis of capital* (New York: Monthly Review Press, 2010), 108.

131 Christopher Flavin, and Gary Gardner, "China, India, and the new world order," in Linda Starke, ed., *State of the World 2006* (New York: W.W. Norton & Company, 2006), 3-23.

132 John Bellamy Foster, "Capitalism and the accumulation of catastrophe," *Monthly Review* 63 (December 2011), 14.

133 James O'Connor, *Natural causes: Essays in ecological Marxism* (New York: The Guilford Press, 1997), 185.

134 James O'Connor, *Natural causes: Essays in ecological Marxism* (New York: The Guilford Press, 1997), 185.

135 James O'Connor, *Natural causes: Essays in ecological Marxism* (New York: The Guilford Press, 1997), 166.

136 James Gustave Speth, *The bridge at the edge of the world: Capitalism, the environment, and crossing from crisis to sustainability* (New Haven: Yale University Press, 2008), 7.

137 David Sawyer, and Seton Stiebert, *Fossil fuels – At what cost?* (2010), http://www.iisd.org/gsi/sites/default/files/ffs_awc_3canprovinces.pdf

138 James Hansen, *Storms of my grandchildren* (New York: Bloomsbury, 2010), 230.

139 Wikipedia, as quoted in Gwynne Dyer, *Climate wars* (Toronto, Random House Canda, 2008).

140 The Royal Society, *Geoengineering the climate: Science, governance and uncertainty* (September 2009), http://royalsociety.org/uploadedFiles/Royal_Society_Content/policy/publications/2009/8693.pdf

141 John Bellamy Foster, "Capitalism and the accumulation of catastrophe," *Monthly Review* 63 (December 2011).

142 James Maitland, as quoted in John Bellamy Foster, and Brett Clark, "The paradox of wealth: Capitalism and ecological destruction" *Monthly Review* 60 (June 2009), 1-18.

143 Herman Daly, as quoted in John Bellamy Foster, and Brett Clark, "The paradox of wealth: Capitalism and ecological destruction" *Monthly Review* 60 (June 2009), 1-18.

[144] Based on a high-fertility scenario in: United Nations, *World population to 2300,* http://www.un.org/esa/population/publications/longrange2/WorldPop2300final.pdf

[145] Simon Caney, "Climate change, human rights, and moral thresholds," in Stephen M. Gardiner, Simon Caney, Dale Jamieson, and Henry Shue, eds., *Climate ethics: Essential readings* (New York: Oxford University Press, 2010), 163-177.

[146] Karl Marx, *Capital: A critique of political economy*, vol.1, (New York: Vintage, 1977), 798.

[147] Wolfgang Lecher, as quoted in André Gorz, Critique of Economic Reason (New York: Verso, 2010), 66.

[148] John Bellamy Foster, Robert W. McChesney, and R. Jamil Jonna, "The global reserve army of labor and the new imperialism," *Monthly Review* 63 (November 2011).

[149] Linda Starke, ed., *Vital signs 2011: The trends that are shaping our future* (New York: W.W. Norton & Company, 2011).

[150] Food and Agricultural Organization of the United Nations, *The state of the world's land and water resources for food and agriculture: Managing systems at risk* (2011), http://www.fao.org/fileadmin/templates/solaw/files/executive_summary/SOLAW_EX_SUMM_WEB_EN.pdf

[151] Linda Starke, ed., *Vital signs 2010: The trends that are shaping our future* (New York: W.W. Norton & Company, 2010).

[152] Lester R. Brown, "Eradicating hunger: A growing challenge," in Linda Starke, ed., *State of the World 2001* (New York: W.W. Norton & Company, 2001), 43-62.

[153] Doug Gurian-Sherman, *Failure to yield: Evaluating the performance of genetically engineering crops* (2009), http://consin.org/view/failure-to-yieldfn.pdf

[154] Maria Alice Garcia, and Miguel A. Altieri, Transgenic crops: Implications for biodiversity and sustainable agriculture (2005), http://agroeco.org/wp-content/uploads/2010/09/garcia-altieri.pdf

[155] Brian Halweil, "Farming in the public interest," in Linda Starke, ed., *State of the World 2002* (New York: W.W. Norton & Company, 2002), 51-74.

[156] William D. Heffernan, "Concentration of ownership and control in agriculture," in Fred Magdoff, John Bellamy Foster, and Frederick H. Buttel, eds., *Hungry for profit*: The agribusiness threat to farmers, food, and the environment (New York: Monthly Review Press, 2000), 65.

[157] Philip McMichael, "The world food crisis in historical perspective," in Fred Magdoff and Brian Tokar, eds., *Agriculture and food in crisis: conflict, resistance, and renewal* (New York: Monthly Review Press, 2010), 51-68.

[158] Albert Bates, and Toby Hemenway, "From agriculture to permaculture," in Linda Starke & Lisa Mastny. eds., *State of the World 2010: Transforming Cultures* (New York: W.W. Norton & Company, 2010), 47-56.

[159] United States Department of Agriculture, *Nutrient Data,* http://www.ars.usda.gov/main/site_main.htm?modecode=12-35-45-00

[160] Institute of Science in Society, *Feeding the world under climate change,* http://www.i-sis.org.uk/FTWUCC.php

[161] Food and Agricultural Organization of the United Nations, *The state of the world's land and water resources for food and agriculture: Managing systems at risk* (2011), http://www.fao.org/fileadmin/templates/solaw/files/executive_summary/SOLAW_EX_SUMM_WEB_EN.pdf

[162] "Land degradation on the rise," *FOANewsroom,* July 2, 2008, http://www.fao.org/newsroom/en/news/2008/1000874/index.html

[163] Peter M. Vitousek, Paul R. Ehrlich, Anne H. Ehrlich, and Pamela A. Matson, "Human appropriation of the products of photosynthesis," in *Bioscience* 36 (June 1986), http://www.rachel.org/files/document/Human_Appropriation_of_the_Products_of_Photosy.pdf

---

[164] Helmut Haberl, Karl-Heinz Erb, and Fridolin Krausmann, *Human appropriation of net primary production (HANPP)* (2007), https://www.ecoeco.org/pdf/2007_march_hanpp.pdf

[165] Mongabay.com, *Calculating deforestation figures for the Amazon,* http://rainforests.mongabay.com/amazon/deforestation_calculations.html

[166] Dale Allen Pfeiffer, *Eating fossil fuels: Oil, food and the coming crisis in agriculture* (Gabriola Island: New Society Publishers, 2006), 11.

[167] United Nations Convention to Combat Desertification, *UNCCD Publications,* http://www.unccd.int/en/resources/publication/Pages/default.aspx

[168] Jared Diamond, *Collapse: How societies choose to fail or succeed* (New York: Penguin Books, 2005).

[169] Philip McMichael, "Feeding the world: Agriculture, development and ecology," in Leo Panitch, and Colin Leys, eds.), *Socialist Register 2007: Coming to terms with nature* (New York: Monthly Review Press, 2007).

[170] Martin Teitel, and Kimberly A. Wilson, *Genetically engineered food: Changing the nature of nature* (Vermont: Park Street Press, 1999), 85.

[171] David McNally, "Slump, austerity and resistance," in Leo Panich, Greg Albo, and Vivek Chibber, eds., *The Crisis and the Left: The Socialist Register 2012* (New York: The Monthly Review Press, 2011): 36-63, 48.

[172] Food and Agricultural Organization of the United Nations, *The state of the world's land and water resources for food and agriculture: Managing systems at risk* (2011), http://www.fao.org/fileadmin/templates/solaw/files/executive_summary/SOLAW_EX_SUMM_WEB_EN.pdf

[173] David McNally, "Slump, austerity and resistance," in Leo Panich, Greg Albo, and Vivek Chibber, eds., *The Crisis and the Left: The Socialist Register 2012* (New York: The Monthly Review Press, 2011): 36-63, 48.

[174] Fred Magdoff, and Brian Tokar, "Agriculture and food in crisis: An overview," in Fred Magdoff and Brian Tokar, eds., *Agriculture and food in crisis: conflict, resistance, and renewal* (New York: Monthly Review Press, 2010), 9-30.

[175] John Bellamy Foster, Brett Clark and Richard York, *The ecological rift: Capitalism's war on the earth* (New York: Monthly Review Press, 2010), 424.

[176] Food and Agricultural Organization of the United Nations, *The state of the world's land and water resources for food and agriculture: Managing systems at risk* (2011), http://www.fao.org/fileadmin/templates/solaw/files/executive_summary/SOLAW_EX_SUMM_WEB_EN.pdf

[177] Institute of Science in Society, *Feeding the world under climate change,* http://www.i-sis.org.uk/FTWUCC.php

[178] Thomas F. Pawlick, *The end of food: How the food industry is destroying our food supply – and what you can do about it* (Vancouver: Greystone Books, 2006), 88.

[179] Dale Allen Pfeiffer, *Eating fossil fuels: Oil, food and the coming crisis in agriculture* (Gabriola Island: New Society Publishers, 2006), 8.

[180] Albert Bates, and Toby Hemenway, "From agriculture to permaculture," in Linda Starke & Lisa Mastny. eds., *State of the World 2010: Transforming Cultures* (New York: W.W. Norton & Company, 2010), 47-56.

[181] Dale Allen Pfeiffer, *Eating fossil fuels: Oil, food and the coming crisis in agriculture* (Gabriola Island: New Society Publishers, 2006), 9.

[182] David Pimentel, "Reducing energy inputs in the agricultural production system," in Fred Magdoff and Brian Tokar, eds., *Agriculture and food in crisis: conflict, resistance, and renewal* (New York: Monthly Review Press, 2010), 241-252.

[183] As quoted in Philip McMichael, "The world food crisis in historical perspective," in Fred Magdoff and Brian Tokar, eds., *Agriculture and food in crisis: conflict, resistance, and renewal* (New York: Monthly Review Press, 2010), 51-68.

[184] David Pimentel, "Reducing energy inputs in the agricultural production system," in Fred Magdoff and Brian Tokar, eds.,

Agriculture and food in crisis: conflict, resistance, and renewal (New York: Monthly Review Press, 2010), 241-252.

185 David Pimentel, "Reducing energy inputs in the agricultural production system," in Fred Magdoff and Brian Tokar, eds., *Agriculture and food in crisis: conflict, resistance, and renewal* (New York: Monthly Review Press, 2010), 241-252.

186 Philip McMichael, "The world food crisis in historical perspective," in Fred Magdoff and Brian Tokar, eds., *Agriculture and food in crisis: conflict, resistance, and renewal* (New York: Monthly Review Press, 2010), 51-68.

187 Max Weber, as quoted in André Gorz, Critique of Economic Reason (New York: Verso, 2010), 37.

188 John Bellamy Foster, "Monopoly Capital and the New Globalization," *Monthly Review* 53 (January 2002).

189 Paul A. Baran, and Paul M. Sweezy, *Monopoly capital: An essay on the American economic and social order* (New York: Monthly Review Press, 1966), 79.

190 Richard Peet, "Contradictions of finance capitalism," *Monthly Review* 63 (December 2011).

191 John Bellamy Foster, and Robert W. McChesney, "What Recovery?" *Monthly Review* 54 (April 2003).

192 Mark Anielski, *The economics of happiness: Building genuine wealth* (Gabriola Island: New Society Publishers), 185.

193 New economics foundation, *Chasing progress: Beyond measuring economic growth* (2004), http://neweconomics.org

194 U.S. National Debt Clock. http://www.brillig.com/debt_clock (Retrieved May 8, 2012)

195 Statistics Brain, Credit card debt statistics, http://www.statisticbrain.com/credit-card-debt-statistics/

196 Fred Moseley, "Goldilocks meets a bear: How bad will the U.S. recession be?" *Monthly Review* 53 (April 2002).

197 John Bellamy Foster, and Robert W. McChesney, "What Recovery?" *Monthly Review* 54 (April 2003).

198 Fred Moseley, "Goldilocks meets a bear: How bad will the U.S. recession be?" *Monthly Review* 53 (April 2002).

---

[199] John Bellamy Foster, and Robert W. McChesney, "What Recovery?" *Monthly Review* 54 (April 2003).

[200] MInqi Li, "After neoliberalism: Empire, social democracy, or socialism?" *Monthly Review* 55 (January 2004).

[201] John Ralston Saul, *The collapse of globalism and the reinvention of the world.* (Toronto: Viking Canada, 2005), 22.

[202] John Ralston Saul, *Voltaire's bastards: The dictatorship of reason in the west* (New York: Penguin Books, 1992), 394.

[203] K.B. Anderson, and & P. Hudis, eds., *The Rosa Luxemburg reader* (New York: Monthly Review Press, 2004).

[204] Paul M. Sweezy, *The theory of capitalist development* (New York: Monthly Review Press, 1970), 227.

[205] Paul M. Sweezy, *The theory of capitalist development* (New York: Monthly Review Press, 1970), 281.

[206] R.H. Tawney, *The acquisitive society* (New York: Dover Publications, Inc., 1920), 37.

[207] Paul M. Sweezy, *The theory of capitalist development* (New York: Monthly Review Press, 1970), 309.

[208] John Bellamy Foster, "Capitalism and the accumulation of catastrophe," *Monthly Review* 63 (December 2011), 12.

[209] John W. Farley, "Petroleum and propaganda: The anatomy of the global warming denial industry," *Monthly Review* 64 (May 2012). Union of Concerned Scientists, A climate of corporate control: How corporations have influenced the U.S dialogue on climate science and policy (May 2012), http://www.ucsusa.org/assets/documents/scientific_integrity/a-climate-of-corporate-control-report.pdf

[210] John Gray, as quoted in Simon Critchley, *The faith of the faithless: Experiments in political theology* (New York: Verso, 2012).

[211] Erik Assadourian, "The rise and fall of consumer cultures," in Linda Starke & Lisa Mastny. eds., *State of the World 2010: Transforming Cultures* (New York: W.W. Norton & Company, 2010), 3-20.

[212] Simon Caney, "Cosmopolitan justice, responsibility, and global climate change," in Stephen M. Gardiner, Simon Caney, Dale Jamieson, and Henry Shue, eds., *Climate ethics: Essential readings* (New York: Oxford University Press, 2010), 122-145.

[213] David Keith, as quoted in Gwynne Dyer, *Climate wars* (Toronto: Random House, 2008).

[214] David Victor, *The politics of fossil-fuel subsidies* (2009), http://www.iisd.org/gsi/sites/default/files/politics_ffs.pdf

[215] Daniel J. Weiss, Jackie Weidman, and Rebecca Leber, Big oil's banner year: Higher prices, record profits, less oil (2012), http://thinkprogress.org/climate/2012/02/08/421061/big-oil-higher-prices-record-profits-less-oil/

[216] Suzanne Goldenberg, "Conservative thinktanks step up attacks against Obama's clean energy strategy," The Guardian, May 8, 2012.

[217] Timothy Morton, "Peak nature," Adbusters 19 (Nov/Dec 2011).

[218] James Hoggan, *Climate cover-up: The crusade to deny global warming* (Vancouver: Greystone Books, 2009), 153.

[219] Eric Pooley, *How much would you pay to save the planet? The American press and economics of climate change* (2008), http://www.hks.harvard.edu/presspol/publications/papers/discussion_papers/d49_pooley.pdf

[220] Eric Pooley, *How much would you pay to save the planet? The American press and economics of climate change* (2008), http://www.hks.harvard.edu/presspol/publications/papers/discussion_papers/d49_pooley.pdf

[221] "The press and the pipeline," *Media Matters*, January 26, 2012 (http://mediamatters.org/)

[222] An interesting short film called "Oil in Eden" created by Pacific Wild is informative (http://www.pacificwild.org/site/press/1285692883.html)

[223] Robert W. McChesney, *The political economy of media: Enduring issues, emerging dilemmas* (New York: Monthly Review Press, 2008).

[224] Robert W. McChesney, *The political economy of media: Enduring issues, emerging dilemmas* (New York: Monthly Review Press, 2008).

[225] Eric Pooley, *How much would you pay to save the planet? The American press and economics of climate change* (2008), http://www.hks.harvard.edu/presspol/publications/papers/discussion_papers/d49_pooley.pdf

[226] David Harvey, *Rebel cities: From the right to the city to the urban revolution* (New York: Verso, 2012), 161.

[227] Lawrence Soley, *Censorship Inc.: The corporate threat to free speech in the United States* (New York: Monthly Review Press, 2002).

[228] Linda Starke, ed., *Vital signs 2010: The trends that are shaping our future* (New York: W.W. Norton & Company, 2010), 75.

[229] Steve Horn, "ALEC model bill behind push to require climate denial instruction in schools," *Desmogblog*, January 26, 2012, http://www.desmogblog.com/alec-model-bill-behind-push-require-climate-denial-instruction-schools

[230] Noam Chomsky, *Making the future: Occupations, interventions, empire and resistance* (San Francisco: City Lights Books, 2012), 112.

[231] Center for Responsive Politics, *Lobbying database*, http://www.opensecrets.org/lobby/index.php

[232] Christine Mahoney, "Lobbying success in the United States and the European Union," *Journal of Public Policy* 75 (2007), 35-56, http://faculty.maxwell.syr.edu/chmahone/Mahoney_JPP_2007.pdf

[233] Matthew D. Hill, G. Wayne Kelly, and Robert A. Van Ness, *Determinants and effects of corporate lobbying*. http://faculty.bus.olemiss.edu/rvanness/Working%20Papers/Lobbying.pdf

[234] Daniel J. Weiss, Jackie Weidman, and Rebecca Leber, Big oil's banner year: Higher prices, record profits, less oil (2012), http://thinkprogress.org/climate/2012/02/08/421061/big-oil-higher-prices-record-profits-less-oil/

[235] Bing Guo, *Lobby or contribute? The impacts of corporate governance on firms' political strategies* (November 24, 2009), http://idea.uab.es/bguo/paper2new.pdf

[236] Michel Foucault, *The Government of Self and Others: Lectures at the Collège de France, 1982-1983* (New York: Palgrave Macmillan, 2010), 184.

[237] Noam Chomsky, *Hopes and prospects* (Chicago: Haymarket Books, 2010), 108.

[238] Noam Chomsky, *Hopes and prospects* (Chicago: Haymarket Books, 2010), 209.

[239] Francis Fox Piven, "The new American poor law," in Leo Panich, Greg Albo, and Vivek Chibber, eds., *The Crisis and the Left: The Socialist Register 2012* (New York: The Monthly Review Press, 2011): 107-124.

[240] Francis Fox Piven, "The new American poor law," in Leo Panich, Greg Albo, and Vivek Chibber, eds., *The Crisis and the Left: The Socialist Register 2012* (New York: The Monthly Review Press, 2011): 107-124.

[241] Al Gore, *The assault on reason* (New York: The Penguin Press, 2007), 75.

[242] Charles Taylor, *The Malaise of Modernity* (Toronto: House of Anansi Press Limited, 1991), 9.

[243] Jon H. Pammett, and Lawrence LeDuc, *Explaining the turnout decline in Canadian federal elections: A new survey of non-voters* (2003), http://www.elections.ca/res/rec/part/tud/TurnoutDecline.pdf

[244] Murray Bookchin, *Urbanization without Cities: The Rise and Decline of Citizenship* (Montreal: Black Rose Press, 1992), 250.

[245] John Berger, "Where are we?" *Harper's Magazine* 306 (March 2003), 13-17.

[246] David Graeber, *Possibilities: Essays on hierarchy, rebellion, and desire* (Oakland: AK Press, 2007), 302.

[247] Jon H. Pammett, and Lawrence LeDuc, *Explaining the turnout decline in Canadian federal elections: A new survey of non-voters* (2003), http://www.elections.ca/res/rec/part/tud/TurnoutDecline.pdf

[248] Herbert Marcuse, *One-dimensional Man* (Boston: Beacon Press, 1964), 7.

[249] David Schweickart, *Against capitalism* (Boulder, Colorado: Westview Press, 1996), 182.

250 As quoted in Peter Marshall, *Demanding the impossible: A history of anarchism* (London: Fontana Press, 1993), 23.

251 Paul M. Sweezy, and Leo Huberman, as quoted in "Notes from the editors," *Monthly Review*, 62 (September 2010) , 9.

252 Robert D. Putnam, *Bowling Alone* (New York: Simon & Schuster, 2001), 21.

253 Robert D. Putnam, *Bowling Alone* (New York: Simon & Schuster, 2001), 357.

254 Robert D. Putnam, *Bowling Alone* (New York: Simon & Schuster, 2001), 288.

255 Chris Mooney, "The science of why we don't believe in science," *Mother Jones* (2011), http://motherjones.com/politics/2011/03/denial-science-chris-mooney

256 Aaron M. McCright, and Riley E. Dunlap, "The politicization of climate change and polarization in the American public's views of global warming, 2001-2010," *The Sociological Quarterly* 52 (2011), 155-194, http://news.msu.edu/media/documents/2011/04/593fe28b-fbc7-4a86-850a-2fe029dbeb41.pdf

257 Lawrence C. Hamilton, *Climate change: Partisanship, understanding, and public opinion* (Spring 2011), http://www.climateaccess.org/sites/default/files/Hamilton_Climate%20Change%20Partisanship,%20Understanding,%20and%20Public%20Opinion.pdf

258 Dan M. Kahan, et. al., *The tragedy of the risk-perception commons: Culture conflict, rationality conflict, and climate change* (2011), http://www.law.upenn.edu/academics/institutes/regulation/papers/Kahan%20Tragedy%20of%20the%20Risk-Perception.pdf

259 Lorrain Hansberry, *A Raison in the sun* (New York: Random House, 1957), 138.

260 Elinor Ostrom, *Governing the commons: The evolution of institutions for collective action*. (Cambridge: Cambridge University Press, 1990), 44.

[261] James Hoggan, *Climate cover-up: The crusade to deny global warming* (Vancouver: Greystone Books, 2009), 222.

[262] Joseph Heath, *Filthy lucre: Economics for people who hate capitalism* (Toronto: Harper Collins Publishers Ltd., 2009), 265.

[263] Waveyard, *Prepare for the dream,* http://www.waveyard.com/

[264] The Playmania, *Ski Dubai,* http://www.theplaymania.com/skidubai

[265] John Bellamy Foster, Brett Clark and Richard York, *The ecological rift: Capitalism's war on the earth* (New York: Monthly Review Press, 2010), 424.

[266] Madeleine Bunting, as quoted in Patrick Curry, *Ecological ethics: An introduction* (Cambridge: Polity Press, 2006), 10.

[267] Hubert Reeves, *Terracide* (Markham, Ontario: Cormorant Books, Inc., 2009), 154.

[268] Ronald Wright, *A short history of progress* (Toronto: House of Anansi, 2004), 125.

[269] James Gustave Speth, *The bridge at the edge of the world: Capitalism, the environment, and crossing from crisis to sustainability* (New Haven: Yale University Press, 2008), 1.

[270] Alternatives to Economic Globalization, as quoted in James Gustave Speth, *The bridge at the edge of the world: Capitalism, the environment, and crossing from crisis to sustainability* (New Haven: Yale University Press, 2008), 173.

[271] James Lovelock, *The vanishing face of Gaia: A final warning* (New York: Penguin Books, 2009), 47.

[272] Patrick Curry, *Ecological ethics: An introduction* (Cambridge: Polity Press, 2006), 12.

[273] Thomas Homer-Dixon, *The ingenuity gap: Can we solve the problems of the future?* (Toronto: Vintage Canada Ltd., 2001), 375.

[274] Neil Evernden, *The natural alien: Humankind and environment,* 2nd ed. (Toronto: University of Toronto Press, 1999), 128.

[275] Noam Chomsky, *Making the future: Occupations, interventions, empire and resistance* (San Francisco: City Lights Books, 2012), 73.

[276] Statcan, *Household access and use of recycling programs*, http://www.statcan.gc.ca/pub/16-002-x/2007001/article/10174-eng.htm#table1

[277] Energy Information Administration, *Methodology for allocating municipal solid waste to biogenic and non-biogenic energy* (May 2007), http://205.254.135.7/totalenergy/data/monthly/pdf/historical/msw.pdf

[278] Natural Resources Canada, *About electricity*, http://www.nrcan.gc.ca/energy/sources/electricity/1387#domestic

CPSIA information can be obtained at www.ICGtesting.com
Printed in the USA
LVOW07s1228251114

415405LV00001B/6/P